AGRICULTURE ISSUES AND POLICIES

U.S. FAMILY FARMS

STRUCTURE, FINANCES, AND AGRICULTURAL PRODUCTION ROLE

AGRICULTURE ISSUES AND POLICIES

Additional books in this series can be found on Nova's website
under the Series tab.

Additional e-books in this series can be found on Nova's website
under the e-book tab.

AGRICULTURE ISSUES AND POLICIES

U.S. FAMILY FARMS

STRUCTURE, FINANCES, AND AGRICULTURAL PRODUCTION ROLE

KATHLEEN PERRY
EDITOR

nova publishers
New York

Library of Congress Cataloging-in-Publication Data

ISBN: 978-1-63483-629-6

Published by Nova Science Publishers, Inc. † New York

Contents

PREFACE

Most U.S. farms are family operations, and even the largest farms are predominantly family-run. Small farms are less profitable than larger farms, on average, and their operator households tend to rely on off-farm income for their livelihood. Generally speaking, farm operator households cannot be characterized as low-income when both farm and off-farm incomes are considered. Small family farms dominate the farm count, but midsize and large-scale family farms account for the bulk of farm production. Information on the different kinds of farms—and the farmers who operate them—is important for understanding the economic wellbeing of farm households and the implications of farm policy. This book proves accurate, detailed, and unbiased information on the structure and finances of U.S. farms, including the relationship of farm size and type to agricultural production, financial performance, sources of farm household income, and the extent of off-farm work. The book provides a sense of the financial position of family farms in general and of different types of family farms.

In: U.S. Family Farms
Editor: Kathleen Perry

ISBN: 978-1-63483-629-6

Chapter 1

STRUCTURE AND FINANCES OF U.S. FARMS: FAMILY FARM REPORT, 2014 EDITION[*]

Robert A. Hoppe

ABSTRACT

Most U.S. farms—97 percent in 2011—are family operations, and even the largest farms are predominantly family-run. Midsize and large-scale family farms account for 8 percent of U.S. farms but 60 percent of the value of production. In contrast, small family farms make up 90 percent of the U.S. farm count but produce a more modest 26-percent share of farm output. Nonfamily farms account for the remaining 3 percent of farms and 15 percent of production. Small farms are less profitable than larger farms, on average, and their operator households tend to rely on off-farm income for their livelihood. Generally speaking, farm operator households cannot be characterized as low-income when both farm and off-farm incomes are considered. Nevertheless, limited-resource farms still exist and account for 11 percent of family farms.

---

[*] This is an edited, reformatted and augmented version of Economic Information Bulletin Number 132, issued by the U.S. Department of Agriculture, Economic Research Service, December 2014.

WHAT IS THE ISSUE?

Broad descriptions of farms based on U.S. averages can mask variation among different sizes and types of farms. Small family farms dominate the farm count, but midsize and large-scale family farms account for the bulk of farm production. Information on the different kinds of farms—and the farmers who operate them—is important for understanding the economic wellbeing of farm households and the implications of farm policy.

To provide information on how U.S. farming is organized, USDA's Economic Research Service (ERS) produces a periodic report documenting the role of family farms in U.S. agricultural production. The *Family Farm Report, 2014 Edition* is the most recent in the series, providing accurate, detailed, and unbiased information on the structure and finances of U.S. farms, including the relationship of farm size and type to agricultural production, financial performance, sources of farm household income, and the extent of off-farm work. The report provides a sense of the financial position of family farms in general and of different types of family farms.

WHAT DID THE STUDY FIND?

Family farms accounted for 97 percent of U.S. farms in 2011. Small family farms alone— those reporting annual gross cash farm income (GCFI) less than $350,000—made up 90 percent of farms (see table for farm types). They also operated 52 percent of the Nation's farmland. In contrast, small farms accounted for a relatively small share of production, 26 percent, although their share of production was much higher for specific commodities. For example, small farms accounted for 56 percent of poultry production, which accounted for the largest share of small farms' production under contract.

Midsize and large-scale family farms together produce the bulk of agricultural output. Large-scale and midsize family farms made up only 8 percent of all U.S. farms in 2011, but they accounted for 60 percent of the value of U.S. agricultural production. Another 3 percent of farms were nonfamily farms, producing 15 percent of U.S. farm output; roughly 85 percent of nonfamily farm output was on farms with GCFI of $1,000,000 or more. Most nonfamily farms (78 percent), however, had GCFI below the $350,000 cutoff used to identify small farms.

Small family farms are more likely to have profitability measures that fall in the critical zone, indicating potential financial problems. About three-fourths of U.S. farms are in the critical zone for rate of return on assets (a value less than 1 percent), and two-thirds are in the critical zone for operating profit margin (a value less than 10 percent). The shares in these critical zones are especially high for farms in the retirement, off-farm occupation, and low-sales categories, tapering off rapidly as farm size (measured by GCFI) increases.

Classifying farms by operator's primary occupation and farm size

Farm type	Operator's primary occupation[1]	Farm size, measured by annual GCFI[2]
Small family farms[3]	Varies	Less than $350,000
Retirement farms	Retired from farming	Less than $350,000
Off-farm occupation farms	Nonfarm	Less than $350,000
Farm-occupation farms:		
Low-sales	Farming	Less than $150,000
Moderate-sales	Farming	$150,000-$349,999
Midsize family farms[3]	Not a criterion	$350,000-$999,999
Large-scale family farms[3]	Not a criterion	$1,000,000 or more
Large farms	Not a criterion	$1,000,000-$4,999,999
Very large farms	Not a criterion	$5,000,000 or more
Nonfamily farms[3]	Not a criterion	Not a criterion

[1]Occupation at which the operator spends 50 percent or more of his or her work time.

[2]Gross cash farm income (GCFI) is the sum of the farm's crop and livestock sales, Government payments, and other farm-related income.

[3]Family farms include any farm where the majority of the business is owned by the operator and relatives of the operator. Nonfamily farms do not meet that requirement.

Small-farm households rely on off-farm income. Given small farms' poor financial performance, why do so many continue to exist? Small-farm households typically receive substantial off-farm income and do not rely primarily on their farms for their livelihood. They often invest in their farm operations with off-farm income. Except for households operating retirement farms, most of their off-farm income is from wage-and-salary jobs or self-employment. Households operating retirement farms typically receive most of their off-farm income from such sources as Social Security, pensions, dividends, interest, and rent.

Farm operator households, generally speaking, cannot be considered low-income. Median household income for only two types of farm households—those operating retirement or low-sales farms—was below the median for all U.S. households in 2011. Nevertheless, the net worth of nearly all households operating retirement or low-sales farms, 96 and 97 percent, respectively, was higher than the median for all U.S. households. Large majorities of households operating other types of farms had both income and net worth above the corresponding medians for all U.S. households.

How Was the Study Conducted?

The 2011 Agricultural Resource Management Survey (ARMS) is the main source of data in the *Family Farm Report, 2014 Edition*. ARMS is an annual survey designed and conducted by ERS and the National Agricultural Statistics Service, another USDA agency. The survey provides detail on farm finances and farm household characteristics that is not available from the census of agriculture. The 2011 ARMS was analyzed rather than the 2012 ARMS, because the 2012 ARMS was reweighted in June 2014 to make it more consistent with the 2012 Census of Agriculture. Using the 2011 ARMS instead results in a more timely release of this report.

Introduction

Farming in the United States is diverse, ranging from very small retirement and residential farms producing little to enterprises with annual sales in the millions of dollars. Farms are operated by individuals on a part-time basis, by multiple generations of a family, and by managers of nonfamily corporations. Some specialize in a single commodity, while others produce a number of commodities.

Broad descriptions of farms based on U.S. averages can be misleading because they mask variation among different sizes and types of farms. Small family farms dominate the farm count, but midsize and large-scale family farms account for the bulk of farm production. Information on the different kinds of farms—and the farmers who operate them—is important for understanding the economic well-being of farm households and the impact of farm policy.

The *Family Farm Report, 2014 Edition* presents comprehensive information about the structure and finances of the diverse types of family farms in the United States. This report—like earlier editions—covers selected topics, such as the number and size of U.S. farms, their commodity specialization, their use of contracts, the characteristics of their operators, their receipts from Government payments, and the finances of farm businesses and the households that operate them. These topics are critical in understanding how U.S. agriculture is organized.

Defining Farms

The diversity of U.S. farms is partly attributable to the official farm definition, which includes farms that are very small in terms of sales of farm products. A farm is currently defined as any place from which $1,000 or more of agricultural products were sold or would normally have been sold during a given year, with Government payments counted as sales. This definition has been in place since 1975, by joint agreement among the U.S. Department of Agriculture (USDA), the Office of Management and Budget, and the U.S. Census Bureau (O'Donoghue et al., 2009, pp. 3-4). Nominal commodity price increases lead to an increasing number of very small farms over time, because less and less production is necessary to meet the $1,000 cutoff as prices increase.

Farms with gross sales less than $1,000 may also be included in the farm count if they might normally have sales high enough to satisfy the sales requirement. If a place does not have $1,000 in sales, a "point system" assigns points—each valued at $1—for acres of various crops and head of livestock to estimate normal or potential sales.[1] "Point farms" are farms with less than $1,000 in sales but with sales, Government payments, and points worth at least $1,000 (O'Donoghue et al. (2009, pp. 3-4). Point farms accounted for 25 percent of U.S. farms in 2011.

The Farm Typology

USDA's Economic Research Service (ERS) developed a farm typology or classification system in 1998—using 1995 farm survey data—to group farms, particularly small family farms, into more homogeneous categories based primarily on annual gross farm sales, the major occupation of the operator, and family/nonfamily ownership of the farm. Use of these more homogeneous groups helps draw a clearer picture of the status of farms in the United States.

ERS updated the typology recently, nearly 15 years after the release of the original version of the classification (Hoppe and MacDonald, 2013). This is the first edition of the *Family Farm Report* to use the updated typology. The update addressed two recent trends: commodity price increases and a shift in production to larger farms. The previous version of the typology defined small farms as those with sales less than $250,000. The small-farm cutoff was increased to $350,000, reflecting a 41-percent increase in commodity prices between 1995 and 2010. To address the shift in production, two new groups were added: family farms with sales from $1 million to $4,999,999 and family farms with sales of $5 million or more. The original typology had less detail at the upper end of the size spectrum, with all family farms with sales of $500,000 or more placed in a single group.

The agency also introduced a technical change in the measurement of farm sales, shifting from gross farm sales to gross cash farm income (GCFI). Compared with gross farm sales, GCFI is a better measure of the size of the farm business because it focuses more on the revenue actually received by the farm. One consequence of the switch to GCFI is an increase in the share of farms with production contracts classified as small from 33 percent to 75 percent (see box, "From Gross Farm Sales to Gross Cash Farm Income").

The updated typology is shown in the box "The Revised Farm Typology." Just as the small-farm cutoff was increased from $250,000 to $350,000 to reflect commodity price increases, the cutoff for low-sales farms increased from $100,000 to $150,000. There now are four groups of small family farms, a midsize group of family farms, and two groups of large-scale family farms (large and very large).

Under the new typology, 90 percent of U.S. farms are classified as small family farms in 2011, similar to the 87-percent share under the previous version of the typology. The shift in the value of production was more substantial, increasing from 15 percent under the previous typology to 26 percent under the updated version. Virtually all of this increase is associated with production contracts. Farms with production contracts typically have small GCFI relative to their gross farm sales, as explained in the box discussing the two measures of farm size. Many of these farms were classified as larger operations in the previous typology (based on gross farm sales) but are reclassified as small in the revised typology (based on GCFI). These reclassified farms have a substantial value of production, mostly under contract.

From Gross Farm Sales to Gross Cash Farm Income

During the update of the farm typology, the measure of farm size was changed from gross farm sales to gross cash farm income (GCFI). GCFI focuses on the revenue received by the farm business and includes the farm's sales of crops and livestock, receipts of Government payments, and other farm-related income. Gross farm sales differs from GCFI by excluding other farm-related income and by including items that are not revenue to the farm—the value of production accruing to share landlords and production contractors, as well as Government payments accruing to landlords.

What's included?

Item	Gross farm sales	Gross cash farm income
Revenue to the farm from:		
Crop and livestock sales	Yes	Yes
Government payments	Yes	Yes
Other farm-related income[1]	No	Yes
Value of production accruing to:		
Share landlords	Yes	No
Contractors	Yes	No
Landlord receipt of Government payments	Yes	No

[1] Receipts from custom work, machine hire, live-stock grazing fees, timber sales, outdoor recreation, production contract fees, etc.

Gross farm sales focuses on the value of products sold or removed from the farm, regardless of who receives payment for the products. For use in the farm typology, GCFI is a better indicator of the size of the farm business than gross farm sales because it focuses on the revenue actually received by the farm business that it can use.

For most farms, GCFI and gross sales are equal.

These farms do not share production or Government payments with landlords or contractors and do not receive other farm-related income.

Gross farm sales, however, can be much larger than GCFI for farms with livestock production contracts. The value of the livestock removed is included in gross farm sales (as the value of production accruing to contractors). Contract growers receive a production contract fee for their services, but the fee is a fraction of the value of livestock removed. For example, a four-house broiler operation typically would generate $1.1 million in gross farm sales, but its GCFI—including production contract fees and other revenue—would only be $175,000. (Farm assets, operator household income, and hours worked on farm by the operator are similar for the broiler operation and other farms with GCFI of $150,000 to $200,000.)

With a $350,000 small-farm cutoff, the four-house broiler operation would be classified as a large farm using gross farm sales to measure farm size but as a small farm using GCFI. Because contract fees are so small, switching the measure of size to GCFI increases the share of farms with production contracts classified as small from 33 percent to 75 percent. (For more information, see Hoppe and MacDonald, 2013, pp. 14-18.)

Farms with production contracts, 2011

Number	43,633
Classified as small:	
Gross farm sales < $350,000	32.5%
GCFI < $350,000	74.7%

Source: USDA, National Agricultural Statistics Service and Economic Research Service, 2011 Agricultural Resource Management Survey.

The Revised Farm Typology

The farm typology focuses on the "family farm," or any farm where the majority of the business is owned by the operator and individuals related to the operator, including relatives who do not live in the operator's household. Family farms are classified by the primary occupation of the operator (including retirement as an occupation) and the size of the farm, as shown in the table below.

Classifying farms by operator's primary occupation and farm size

Farm type	Operator's primary occupation[1]	Farm size, measured by annual GCFI[2]
Small family farms[3]	Varies	Less than $350,000
Retirement farms	Retired from farming	Less than $350,000
Off-farm occupation farms[4]	Nonfarm	Less than $350,000
Farm-occupation farms:		
Low-sales	Farming	Less than $150,000
Moderate-sales	Farming	$150,000-$349,999
Midsize family farms[3]	Not a criterion	$350,000-$999,999
Large-scale family farms[3]	Not a criterion	$1,000,000 or more
Large farms	Not a criterion	$1,000,000-$4,999,999
Very large farms	Not a criterion	$5,000,000 or more
Nonfamily farms[3]	Not a criterion	Not a criterion

[1]Occupation at which the operator spends 50 percent or more of his or her work time.

[2]Gross cash farm income (GCFI) is the sum of the farm's crop and livestock sales, Government payments, and other farm-related income.

[3]Family farms include any farm where the majority of the business is owned by the operator and relatives of the operator. Nonfamily farms do not meet that requirement.

[4]Includes a small number of farms—13 percent of the group in 2011—whose operators are not currently in the paid workforce.

Data Sources

The Agricultural Resource Management Survey (ARMS), an annual farm survey, is the main source of data in the *Family Farm Report, 2014 Edition*. ARMS is jointly designed and conducted by ERS and the National Agricultural Statistics Service (NASS), another USDA agency. For more information about ARMS, see *ARMS Farm Financial and Crop Production Practices* at: http://www. ers.usda.gov/data-products/arms-farm-financial-and-crop-production-practices.aspx. Differences between estimates from ARMS generally are stressed in this report only if they are significantly different at the 95 percent level or higher.

ARMS was selected as the main data source for the report because of its rich detail on farm finances and operator household characteristics, detail that is not available from the census of agriculture. Information collected by the survey is critical in assessing the financial strength of farms, the economic well-being of farm households, and the impact of farm policy. The 2011 ARMS was analyzed rather than the 2012 ARMS, because the 2012 ARMS was reweighted to make it more consistent with the 2012 Census of Agriculture in June 2014. The final 2012 ARMS dataset— incorporating the new weights and other revisions—was released in November 2014. Effects of the reweight, however, were minor, and using the 2011 ARMS rather than waiting for the final 2012 ARMS had little impact on the estimates and relationships reported here.[2]

The report also draws on estimates of productivity from ERS, annual estimates of the number of farms from NASS, labor force data from the U.S. Department of Labor's (USDL) Bureau of Labor Statistics (BLS), and Federal crop insurance data from USDA's Risk Management Agency (RMA). These additional sources of data are particularly useful when following trends over long periods of time.

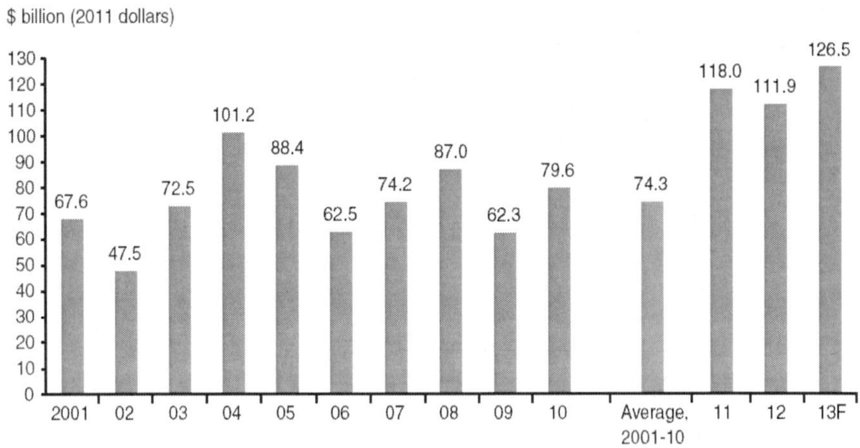

Note: Deflated with the gross domestic product (GDP) chain-type price index.Source: USDA, Economic Research Service, U.S. and State Farm Income and Wealth Statistics (the farm sector accounts), www.ers.usda.gov/data-products/farm-income-and-wealth-statistics.aspx.

Figure 1. Real net farm income, 2001-13Beginning in 2011, net farm income was 50 to 70 percent higher than the average for 2001 to 2010.

The Agricultural Situation in 2011

This report depicts farm structure and financial status as of 2011. That was an above-average year for U.S. farming—as measured by ERS farm-sector income estimates (USDA/ERS, 2014)—reflecting high cash receipts from both crops and livestock. Net farm income was $118 billion in 2011 (Figure 1), 48 percent higher than the previous year and 59 percent higher than the average for the previous 10 years, measured in 2011 constant dollars, using the gross domestic product (GDP) chain-type price index to adjust for price changes. Furthermore, it remained above $100 billion in 2012 and 2013.

U.S. FARMS: NUMBERS, SIZE, AND TENURE

In the 1930s, two important longrun trends began that affected the number of U.S. farms. First, nonagricultural employment resumed growing after the 1933 low point of the Great Depression (Figure 2). Second, farm productivity began to increase steadily (Figure 3), starting about 1937 (Cochrane, 1993, pp. 360-363). Productivity growth exceeded the growth in demand for agricultural products, which led to excess capacity in agriculture, farm consolidation, and farm operators and laborers leaving farming to work in the growing nonfarm economy. The number of farms dropped sharply after a peak of 6.8 million in 1935 to about 3 million in the late 1960s. The decline in farm numbers slowed in the 1970s and nearly stopped in the 1990s. By 2013, about 2.1 million farms remained. Changes in land operated by farms were less dramatic than changes in the number of farms. Land operated by farms peaked at 1.2 billion acres in the early 1950s and then gradually declined to 914 million acres in 2013. Total cropland—a component of farmland operated—has remained relatively constant since World War II but with some small declines since the 1980s (Nickerson et al., 2011, pp 11-13). As farms exited agriculture, most of their land was absorbed—through purchase or rental— by other farms.

Share of Farms, Production, and Farmland

In 2011, 97 percent of U.S. farms were family farms (Figure 4). The remaining 3 percent were non- family farms, which produced 15 percent of the value of U.S. agricultural output. Two features of family farms stand out. First, there are many small family farms (those reporting less than $350,000 in

GCFI), making up 90 percent of all U.S. farms and operating 52 percent of the farmland. Second, most production—60 percent—occurs on the 8 percent of farms classified as midsize or large-scale family farms. As shown in Figure 5, these two groups of family farms dominate U.S. production of cotton (83 percent of production), cash grain (71 percent), amd hogs (66 percent).

Nevertheless, small farms' 26-percent share of production is practically the same share as for midsize farms. Small farms' share of production is substantially higher than 26 percent for specific commodities: 56 percent for poultry, 51 percent for hay, 48 percent for other livestock,[3] and 32 percent for beef. The largest share of small-farm production occurs among moderate-sales farms (GCFI of $150,000-$349,999), which account for 12 percent of total U.S. production (Table 1).

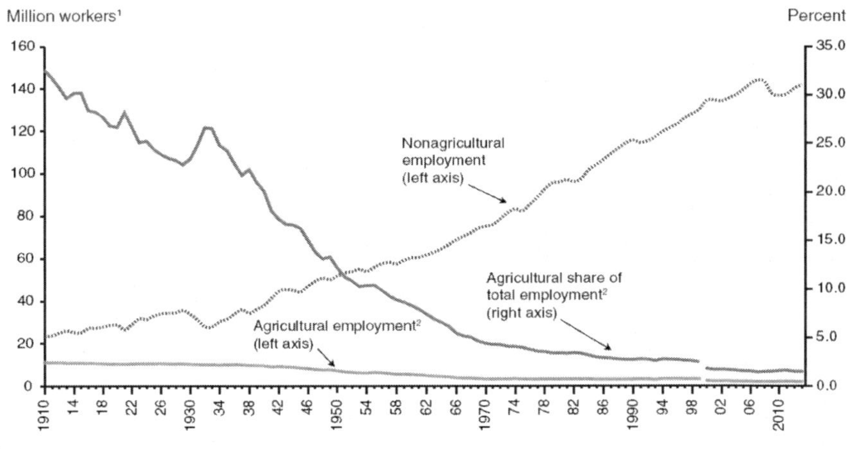

[1] Persons at least 14 years old prior to 1947; persons at least 16 years old in 1947 and later years.

[2] The breaks in the lines indicate a definition change. From 2000 onward, estimates of agricultural employment actually are for "agricultural and related industries." Veterinary and landscaping services were removed from agricultural employment while forestry, fishing, hunting, and trapping were added (U.S. Dept. of Labor, 2003, p. 20). This resulted in a net decrease of about 800,000 "agricultural" employees between 1999 and 2000 and reduced the agricultural share of total employment from 2.5 percent to 1.8 percent.

Source: USDA, Economic Research Service, compiled from U.S. Department of Labor, Bureau of Labor Statistics data in U.S. Executive Office of the President, 2014, pp. 378-379; and U.S. Census Bureau, 1975, p. 12.

Figure 2. Agricultural and nonagricultural employment, 1910-2013Agriculture's share of total U.S. employment fell during the 20th century.

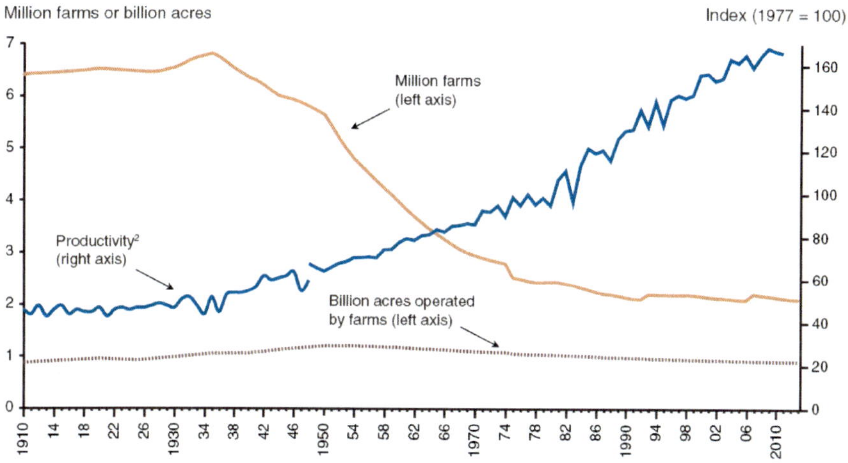

[1] Total factor productivity, or farm output per unit of total factor input (labor, capital, and all other inputs used in production). For more information, see Fuglie et al. (2007).

[2] The break in the productivity line reflects the introduction of new methodology beginning with the 1948 estimate. For more information, see Ahearn et al. (1998, pp. 15-21).

Source: USDA, Economic Research Service, compiled from USDA, National Agricultural Statistics Service annual estimates of the number of farms and acres operated (http://quickstats.nass.usda.gov/) and from ERS estimates of farm productivity (www.ers.usda.gov/data-products/agricultural-productivity-in-the-us.aspx). Acres operated prior to 1950 are from censuses of agriculture for various years, with interpolations between census years. ERS productivity indices prior to 1948 came from Johnson (1990).

Figure 3. Number of farms, farm productivity1 and acres operated, 1910-2013*The number of farms declined as productivity increased.*

Farm Size

Farm size can be measured by the level of GCFI, acres operated, or annual hours of labor used. For our purposes, farm size is best measured by GCFI, which is a better measure of the farm's economic activity than either acreage operated or labor used. Farmland can be of different quality, can be farmed at different levels of intensity, and can produce a variety of commodities. As a result, revenue generated per acre ranges widely across farms. Similarly, labor used depends on the specialization of the farm. For example, large-scale cash grain farms use more machinery and land but less labor than comparable specialty crop farms (Hoppe et al., 2008, pp. 28 and 30). GCFI measures total revenue generated by the farm, rather than the level of one input (land or labor). Nevertheless, land and labor are important inputs and are also examined.

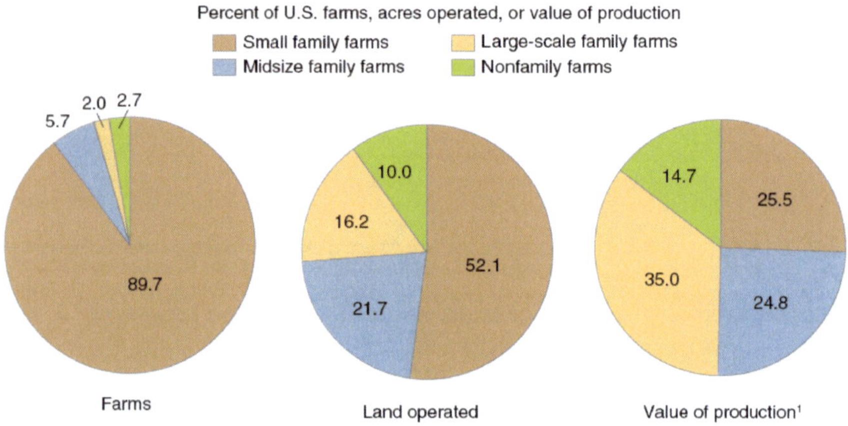

[1] The value of production measures the value of commodities produced in a given year, without the effects of inventory change. It is calculated by multiplying the quantity of each commodity produced by the price of the commodity.

Source: USDA, National Agricultural Statistics Service and Economic Research Service, 2011 Agricultural Resource Management Survey, Phase III.

Figure 4. Share of total farms, acres operated, and value of production, by farm type, 2011Small family farms account for 90 percent of U.S. farms and 52 percent of acres operated, but only 26 percent of production.

Gross Cash Farm Income

Differences in farm size—as measured by GCFI—help drive the distribution of agricultural production. The 1.3 million retirement and off-farm occupation farms account for only 7 percent of production because most of these farms are very small (Table 1). Roughly 70 percent of the farms in each group have GCFI less than $10,000. Similarly, 30 percent of farms in each of these two groups have GCFI less than $1,000. Ninety-nine percent of farms with GCFI less than $1,000 also have gross farm sales less than $1,000 (point farms), as one would expect.[4]

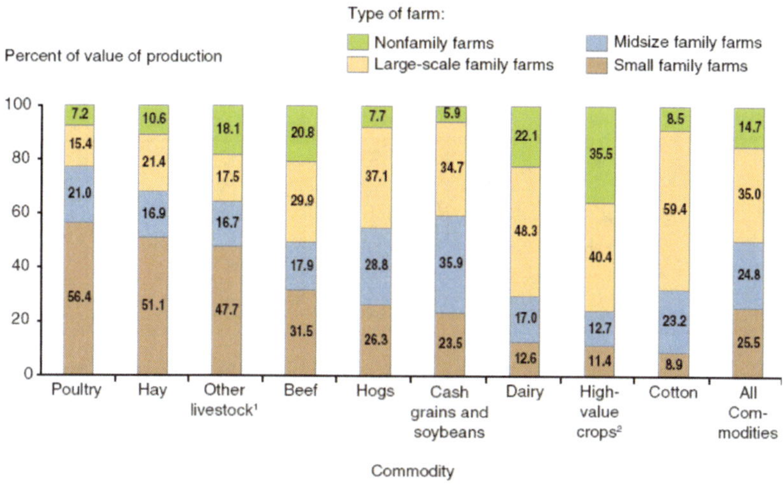

Note: The value of production measures the value of commodities produced in a given year, without the effects of inventory change. It is calculated by multiplying the quantity of each commodity produced by the price of the commodity.1Sheep, lambs, wool, goats, goats' milk, mohair, horses, ponies, mules, donkeys, bees, honey, aquaculture, mink, rabbits, other fur-bearing animals, bison, deer, elk, llamas, etc. 2Vegetables, fruits/tree nuts, and nursery/greenhouse products.Source: USDA, National Agricultural Statistics Service and Economic Research Service, 2011 Agricultural Resource Management Survey, Phase III.

Figure 5. Distribution of the value of production for selected commodities, 2011. *Small farms produce a substantial share of some commodities.*

Nonfamily farms are also concentrated in the lower GCFI classes. Thirty-seven percent have GCFI less than $10,000, and 78 percent have GCFI less than $350,000, the cutoff used to identify small family farms. Only 11 percent have GCFI in excess of $1 million. The only criterion necessary to be classified as a nonfamily farm by the ERS definition is that the operator and the operator's relatives do not own a majority of the business; size is not a criterion. For the most part, nonfamily farms are not large farms operated by publicly held corporations trading on a stock exchange. Only 17 percent of nonfamily farms are organized as corporations, and most of them (94 percent) report no more than 10 stockholders, the original maximum number of stockholders for S-corporations. Nonfamily farms account for 15 percent of total U.S. farm production, but 85 percent of nonfamily farms' production occurs on farms with GCFI of $1 million or more. (See "Appendix I: Large and Small Nonfamily Farms" for more information.)

Table 1. Selected farm characteristics by farm type, 2011

Item	Small family farms				Midsize family farms	Large-scale family farms		Nonfamily farms	All farms
	Retire-ment	Off-farm occupation	Farming-occupation						
			Low-sales	Moderate-sales		Large	Very large		
				Number					
Total farms	353,922	909,872	567,214	118,253	123,009	38,541	3,857	58,175	2,172,843
				Percent of U.S. total					
Distribution of:									
Farms	16.3	41.9	26.1	5.4	5.7	1.8	0.2	2.7	100.0
Value of production[1]	1.5	5.1	6.8	12.0	24.8	23.7	11.3	14.7	100.0
				Percent of group					
GCFI class:									
Less than $1,000	29.0	30.5	15.7	na	na	na	na	17.9	22.0
$1,000-$9,999	43.2	40.3	30.0	na	na	na	na	18.6	32.2
$10,000-$99,999	24.6	25.5	44.9	na	na	na	na	28.6	27.2
$100,000 -$149,999	1.5	1.6	9.4	na	na	na	na	7.1	3.6
$150,000-$349,999	1.8	2.1	na	100.0	na	na	na	5.8	6.8
$350,000-$499,999	na	na	na	na	42.8	na	na	5.0	2.6
$500,000-$999,999	na	na	na	na	57.2	na	na	6.2	3.4
$1,000,000-$4,999,999	na	na	na	na	na	100.0	na	8.1	2.0
$5,000,000-$9,999,999	na	na	na	na	na	na	66.6	1.3	0.2
$10,000,000 or more	na	na	na	na	na	na	33.4	1.5	0.1

Item	Small family farms				Midsize family farms	Large-scale family farms		Nonfamily farms	All farms
	Retire-ment	Off-farm occupation	Farming-occupation						
			Low-sales	Moderate-sales		Large	Very large		
				Acres per farm					
Land operated:									
Median	68	50	92	427	898	2,035	2,480	143	83
Mean	166	145	279	1,022	1,587	3,309	4,927	1,547	415
			Annual person equivalents of labor per farm						
Average person equivalents of labor[2,3]	0.664	0.690	1.180	2.592	3.474	8.060	38.638	4.930	1.386
				Percent of total hours					
Share of hours worked by:[4]									
Principal operator[3]	57.2	61.1	65.2	55.3	40.7	18.1	3.7	14.2	46.5
Spouse[3]	16.4	20.8	16.5	14.3	9.6	4.0	1.0	2.2	12.8
Hired labor	16.0	7.8	8.6	14.9	33.9	57.6	74.4	52.1	25.7
Contract labor	1.6	1.5	1.8	3.4	4.3	10.6	18.1	22.7	6.0
				Percent of group					
Tenure:									
Full owner	87.0	73.5	64.9	31.6	15.6	13.0	23.4	59.2	66.4
Part owner	11.6	22.2	29.0	54.7	72.7	70.6	56.5	26.7	27.9
Tenant	1.4	4 3	6.1	13.7	11.7	16.3	20.2	14.1	5.7

Table 1. (Continued)

Item	Small family farms				Midsize family farms	Large-scale family farms		Nonfamily farms	All farms
	Retire-ment	Off-farm occupation	Farming-occupation						
			Low-sales	Moderate-sales		Large	Very large		
	Percent of land operated								
Share of land rented:									
Part owners	46.0	56.4	49.3	51.1	61.9	64.5	54.0	31.5	55.1

GCFI = Gross cash farm income. na = Not applicable.

[1]The value of production measures the value of commodities produced in a given year, without the effects of inventory change. It is calculated by multiplying the quantity of each commodity produced by the price of the commodity.

[2]One annual person equivalent equals 2,000 hours of labor, or 50 weeks per year times 40 hours per week.

[3]Includes paid and unpaid hours.

[4]Shares worked by other operators and unpaid workers are not shown separately.

Source: USDA, National Agricultural Statistics Service and Economic Research Service, 2011 Agricultural Resource Management Survey, Phase III.

Acres Operated

Retirement and off-farm occupation farms typically operate small acreages; while true whether examining the mean or median, it is particularly evident for median acreage operated by each farm group. For the remaining family farms, acres per farm—both average and median—increase consistently as GCFI goes from less than $150,000 for low-sales farms to at least $5 million for very large farms. Median acres operated range from less than 100 acres for retirement, off-farm occupation, and low-sales farms to nearly 2,500 acres for very large farms (Table 1).

A few high-acreage farms may raise the average well above the acreage operated on most farms. Median acres operated—the midpoint of the distribution of farms by acres operated—avoids this problem. For example, consider the average and median acres per nonfamily farm. The high average acreage for nonfamily farms (1,547 acres) reflects a small share of farms in the group with very large acreages. In contrast, the median for this group is less than one-tenth as large—143 acres— which is consistent with the 78-percent share of nonfamily farms with GCFI less than $350,000.

Labor Hours

Farm operators and their families provide much of the labor—as well as the management—used in farming. Some farms may also have partners unrelated to the operator who serve as additional operators. As farm size increases, the ability of the operator's family to meet the labor requirements of the farm business diminishes, and the farm may use hired or contract farm labor. ARMS data can be used to examine sources of farm labor in detail (see box "Estimating the Sources of Farm Labor).

The person who acts as the employer determines the difference between hired and contract labor (O'Donoghue et al., 2011, p. 5). Farm operations pay hired workers directly, and the worker is a farm employee. In contrast, farm operations pay labor contractors for the performance of specific tasks, such as fruit or vegetable picking. The contract laborers are employees of the contractor, not the farm business.

Estimating the Sources of Farm Labor

Different versions of the Agricultural Resource Management Survey (ARMS) are conducted each year to collect information useful for specific purposes.

All five versions of the 2011 ARMS collected the number of hours worked on farms by the principal operator, the spouse of the principal operator, other operators, and unpaid workers. (ARMS does not differentiate between operators' management and labor hours.) Version 1 of the survey also collected the number of the hours worked by hired laborers. Hours of hired labor on the other versions were estimated by dividing cash wages for hired labor by the region- or State-specific wage rate for farm labor. No versions of the survey collected hours of contract labor, so an estimate was made by dividing contract labor expense by the wage rate. Estimates of wage rates came from *Farm Labor*, published by the National Agricultural Statistics Service (USDA/NASS, 2013b). The report provides wage rates for California, Florida, and 15 regions containing 2 or more States.

One measure of annual labor use is the "person equivalent," defined here as 2,000 hours, or 40 hours of work per week for 50 weeks per year. Off-farm occupation and retirement farms use the least labor, about two-thirds of a person equivalent. Labor use increases to 1-person equivalent for low-sales farms and grows with GCFI to 39-person equivalents for very large farms. Nonfamily farms use 5-person equivalents, on average. That estimate, however, reflects heavy labor use by relatively few farms. Only 12 percent of nonfamily farms use more than 5-person equivalents of labor, while 51 percent use less than 1.

The operator and spouse provide between 70 and 80 percent of the labor on small family farms. Among the remaining family farms, the operator/spouse share declines from 50 percent for midsize farms to 5 percent for very large farms.

This reflects the fact that as farms increase in size, they more often require more labor than the family can provide. In contrast, hired labor increases from 34 percent of the labor needs for midsize farms to 74 percent for very large farms. Contract labor also accounts for roughly a fifth of the work hours on very large and nonfamily farms. Farms specializing in high-value crops use about two-thirds of the contract labor used by the sector; very large family farms and nonfamily farms produce more than half of those crops (not shown in table).

Land Tenure

Land tenure is the ownership status of farmland. Farm operations can be sorted into three categories of tenure:

- **Full owner**—the operation owns all the land farmed.
- **Part owner**—the operation owns some of the land operated and rents the rest.
- **Tenant**—the operation rents all the land operated. Operations that own only a small portion of the land they operate (less than 1 percent) are also considered to be tenants in the ARMS data.

Land leasing is a way of gaining access to additional land as well as a way for beginning farmers to enter agriculture. Farm operations often expand by renting land to avoid debt and the risks associated with ownership and to be able to respond more quickly to changing market conditions. As a result, the share of farms classified as part owners or tenants tends to increase with farm size (Table 1). As a group, part owners rent about half the land they operate.

Large majorities of retirement, off-farm occupation, low-sales, and nonfamily farms are full owners. Leasing is most common among family farms with GCFI of at least $150,000 (moderate-sales, midsize, large, and very large farms). Between 55 and 73 percent of the farms in each of these groups are part owners, and an additional 12 to 20 percent are tenants.

Specialization and Diversification

Over the past century, crop and livestock production largely separated from each other as farmers specialized in the production of a few commodities (MacDonald et al., 2013, pp. 31-33). For example, 75 to 90 percent of farms had chickens, milk cows, or hogs in 1900, but by 2010 less than 10 percent of farms produced those livestock, generally specializing in one species and relying heavily on purchased feeds. More than 80 percent of farms produced corn in 1900, largely to feed their own livestock. By 2010, only one-sixth of farms produced corn, generally specialized crop farms. Specialization allows farmers to capture some efficiencies of scale, but also subjects them to greater market risks as well as production risks from pests and diseases.

Diversification—producing several commodities—can help mitigate these risks, but can also lower resource-use efficiency.

In this report, a farm is said to specialize in a particular crop or livestock commodity if that commodity accounts for at least 50 percent of its production. Diversification is measured by the number of commodities a farm produces. Note that a farm can receive the bulk of its production from one commodity, such as corn, but still produce additional commodities. An example would be a farm specializing in corn but also producing soybeans and hay.

Specialization

Beef cattle is a common specialization among small family farms, accounting for roughly one-fourth to one-third of retirement, off-farm occupation, and low-sales farms (Table 2). Small farms as a whole also account for a substantial share of the total value of beef production—32 percent (see Figure 5)—as mentioned earlier. There actually are three phases in the commercial production of beef cattle (McBride and Mathews, 2011, pp. 5-7; Cash, 2002, p. 21):

- **Cow-calf**—this stage involves maintenance of cows and the production of weaned calves at 6 to 9 months old, when they weigh between 400 and 700 pounds.
- **Stocker**—the calves gain another 200 to 400 pounds on forages—pasture, hay, and crop residues—and grains over a 3- to 8-month period.
- **Feedlot**—the calves are finished on a combination of forages and grain to be sent to slaughter at 1,000 to 1,500 pounds.

Roughly half of beef enterprises focus on the cow-calf phase and the rest participate in two or all three phases. Enterprises specializing in the cow-calf phase are typically found on small farms. Farms with cow-calf enterprises average 57 weaned calves per year and most of them (77 percent) produce hay. Much smaller shares report producing corn (6 percent), soybeans (5 percent) or small grains (10 percent).

Table 2. Farm specialization and diversification by farm type, 2011

Item	Small family farms				Midsize family farms	Large-scale family farms		Nonfamily farms	All farms
			Farming-occupation						
	Retirement	Off-farm occupation	Low-sales	Moderate-sales		Large	Very large		
					Number				
Total farms	353,922	909,872	567,214	118,253	123,009	38,541	3,857	58,175	2,172,843
					Percent				
Commodity specialization:[1]									
Cash grain[2]	6.8	9.3	11.5	40.5	54.5	48.1	24.3	15.7	14.6
Other field crops[3]	42.1	22.3	17.7	6.8	7.7	10.9	4.7	27.3	22.6
High-value crops[4]	4.5	5.5	8.2	8.4	8.2	12.3	21.3	15.7	6.8
Beef	23.6	34.7	35.5	17.2	12.1	11.4	10.4	23.4	30.1
Hogs	0.2	0.7	0.5	1.9	2.8	4.7	3.3	1.4	0.8
Dairy	0.1	0.1	2.9	14.6	10.7	7.4	32.2	2.9	2.5
Poultry	1.8	1.6	3.4	7.0	2.2	2.8	3.1	1.1	2.4
Other livestock[5]	20.9	25.7	20.4	3.6	1.7	2.5	0.7	12.6	20.2
					Number				
Average number of commodites[6]	0.9	1.3	1.7	3.1	3.3	3.4	3.7	1.5	1.6
					Percent				
Number of commodities:[6]									
None	40.1	23.0	16.7	1.1	1.0	0.1	0.0	21.9	21.2
One	39.3	42.7	36.1	19.3	13.0	13.9	19.3	38.2	36.8

Table 2. (Continued)

Item	Small family farms				Midsize family farms	Large-scale family farms		Nonfamily farms	All farms
			Farming-occupation						
	Retirement	Off-farm occupation	Low-sales	Moderate-sales		Large	Very large		
Two	16.2	24.8	29.1	25.1	25.9	21.8	10.5	23.9	24.5
Three	2.6	6.3	9.1	19.4	19.8	23.0	12.5	7.6	8.2
Four or more	1.7	3.3	9.0	35.1	40.3	41.1	57.7	8.5	9.3

[1] Commodity or commodity group that accounts for at least half of the farm's value of production.

[2] Includes wheat, corn, soybeans, grain sorghum, rice, and general cash grains where no single cash grain accounts for the majority of production.

[3] Tobacco, peanuts, cotton, sugar beets, sugar cane, corn for silage, sorghum for silage, hay, canola, and general crops, where no single crop accounts for the majority of production. Also includes farms with all cropland in land-retirement programs.

[4] Vegetables, fruits/tree nuts, and nursery/greenhouse products.

[5] Includes sheep, lambs, wool, goats, goats' milk, mohair, horses, ponies, mules, donkeys, bees, honey, aquaculture, mink, rabbits, other fur-bearing animals, bison, deer, elk, llamas, etc. Also includes farms where no single livestock species accounts for the majority of production.

[6] Based on 27 commodities or commodity groups: barley, oats, wheat, corn for grain, corn silage, soybeans, sorghum for grain, sorghum silage, canola, fruits, vegetables, nursery products, peanuts, sugar cane, sugar beets, rice, potatoes, cotton, tobacco, hay, other crops, cattle, hogs, dairy, poultry, and other livestock.

Source: USDA, National Agricultural Statistics Service and Economic Research Service, 2011 Agricultural Resource Management Survey, Phase III.

Small feedlots exist, but larger operations account for most of the beef shipped directly from feedlots to the slaughter market. The 2012 Census of Agriculture, released by USDA/NASS in May 2014, reported that feedlots selling fewer than 50 head of finished cattle accounted for 50 percent of the 33,880 feedlots in the United States but only 2 percent of the total head of finished cattle (USDA/ NASS, 2014, p. 19). In contrast, 77 percent of finished cattle came from 607 specialized farms that sold 5,000 head or more.

Cattle enterprises offer advantages to operators of small farms. Cattle are less labor-intensive than many other enterprises (except during calving), which may be attractive to an operator who is retired or who holds a full-time job off the farm (Cash, 2002, p. 21). Cattle enterprises also tend to be low-cost, which limits cash requirements.

Other Field Crops and Other Livestock

Two other specializations are common among retirement, off-farm occupation, and low-sales farms (Table 2). Eighteen to 42 percent of the three groups specialize in other field crops, which also includes farms with all their crop acres in land-retirement programs: the USDA Conservation Reserve Program, the Conservation Reserve Enhancement Program, and the Wetlands Reserve Program. Another 20 to 26 percent of each group specializes in other livestock, mostly grazing livestock other than cattle (horses, sheep, and goats).[5]

Grain, Dairy, and Poultry

The grain, dairy, and poultry specializations are most common among moderate-sales, midsize, and large-scale farms. Farms specializing in cash grains account for 24 to 55 percent of these groups of farms. Dairy farms make up 15 and 32 percent of moderate-sales and very large farms, respectively (versus 3 percent of farms in general). Seven percent of moderate-sales farms specialize in poultry— at least double the rate for any other group—which helps explain small farms' 56-percent share of poultry production (see Figure 5).

High-Value Crops

Specialization in high-value crops is most common among large-scale family farms and nonfamily farms, which account for 12 to 21 percent of the farms in these groups (Table 2). These groups together account for three-fourths of the production of high-value crops (see Figure 5). No more than 8 percent of the small or midsize farm types specialize in high-value crops.

High-value crops can generate large sales per acre, but they can require much more labor than most other commodities grown by small-farm operators and they may require more marketing expertise.

Diversification

Many small family farms specialize in a single commodity or produce nothing at all; retirement farms were the most likely to report no commodities in 2011 (Table 2). Farms with no production in 2011 could satisfy the USDA farm definition through a combination of sales of stored commodities produced in previous years, points from the point system, and Government payments. Family farms become more diversified as their size increases. Moderate-sales, midsize, and large-scale farms average three to four commodities per farm, while smaller family farms averaged one or two commodities. Between 35 percent and 58 percent of the four family-farm categories with GCFI of $150,000 or more produced four or more commodities, compared with less than 10 percent of smaller family farms.

CONTRACTING

A contract is a legal agreement between a farm operator and another person or firm (the contractor) to produce a specific type, quantity, and quality of agricultural commodity. ERS identifies two types of contracts: marketing contracts and production contracts (see box, "Types of Contracts"). The share of agricultural production under contract increased from 11 percent in 1969 to 40 percent in 2011. Contracts are especially important for particular commodities, accounting for much of the production of poultry and eggs, hogs, milk, sugar beets, and tobacco (MacDonald and Korb, 2011, 8-14).

Contracts can potentially provide benefits to both producers and contractors (MacDonald and Korb, 2011, pp. 2-6). Farmers get a guaranteed outlet for their production with known compensation, while contractors get an assured supply of commodities with specified characteristics, delivered in a timely manner. Contracts may also bring new risks for farmers. As an example, farmers may produce crops or livestock that do not meet the quality or quantity standards of their contracts, which could force them to fulfill their contracts with open market purchases.

Contracting by Small Farms

Introducing the revised typology increased the share of small-farm production under contract. Under the revised typology, 38 percent of all small-farm production was under contract in 2011, or more than double the 17-percent share had the original typology been used to classify small farms. The increase resulted from shifting to GCFI as the measure of farm size, which moved about 18,500 farms with production contracts—and their substantial production—into small-farm categories. Which small farms use contracts the most? Relatively small shares of retirement and off-farm occupation farms have contracts, 3 and 5 percent, respectively (Table 3). Somewhat more low-sales farms use contracts (9 percent). Moderate-sales farms, however, use contracts the most, with 40 percent reporting one or more contracts. Compared with other small farms, moderate-sales farms also have a larger share of their total production under contract (47 percent), similar to the shares for very large farms (50 percent) and nonfamily farms (45 percent). Poultry ranks first in the share of contract production among small farms—regardless of typology group—with poultry's share of each group's total production under contract ranging from 50 to 64 percent.

Types of Contracts

Two types of contracts are identified in the Agricultural Resource Management Survey (ARMS):

- Marketing contracts. Ownership of the commodity remains with the farmer during production. The contract sets a price (or a pricing formula), product quantities and qualities, and a delivery schedule. Contractor involvement in production is minimal, and the farmer provides all the inputs. For crops, the contract is finalized before harvest. For livestock, the contract is finalized before the transfer.
- Production contracts. The contractor usually owns the commodity during production, and the farmer is paid a fee for services rendered. The contract specifies farmer and contractor responsibilities for inputs and practices. The contractor often provides specific inputs and services, production guidelines, and technical advice. In livestock contracts, for example, contractors typically provide feed, veterinary services, transportation, and young animals.

Source: MacDonald and Korb, 2011, pp. 1-2.

Table 3. Farms with contracts and production under contract by farm type, 2011

Item	Small family farms				Midsize family farms	Large-scale family farms		Nonfamily farms	All farms
		Off-farm occupa-tion	Farming-occupation						
	Retire-ment		Low-sales	Moderate-sales		Large	Very large		
	Number								
Total farms	353,922	909,872	567,214	118,253	123,009	38,541	3,857	58,175	2,172,843
	Percent of group's farms or group's total production								
Farms with contracts1	2.5	4.5	8.8	40.0	54.0	63.5	61.2	19.5	11.6
Value of production under contract2,3	26.6	28.4	31.7	46.8	36.6	39.4	50.3	45.2	40.4
	Commodity (and commodity's share of group's production under contract)4								
Top commodities produced under contract5									
First	Poultry (64)	Poultry (50)	Poultry (61)	Poultry (52)	Grain (35)	Grain (38)	Dairy (38)	H-v6 (33)	Grain (24)
Second	Grain (16)	Grain (25)	Grain (11)	Hogs (20)	Poultry (21)	Hogs (12)	H-v6(21)	Dairy (23)	Poultry (20)
Third	Hogs (9)	Hogs (11)	Hogs (9)	Grain (12)	Hogs (15)	H-v6(11)	Cattle/grain7 (15)	Cattle (21)	Dairy (15)
	Percent of U.S. total								
Farms with contracts1	3.5	16.2	19.8	18.8	26.4	9.7	0.9	4.5	100.0
Value of production3	1.5	5.1	6.8	12.0	24.8	23.7	11.3	14.7	100.0
Under contract2	1.0	3.6	5.3	14.0	22.5	23.1	14.1	16.5	100.0
Not under contract	1.9	6.2	7.8	10.7	26.4	24.1	9.4	13.5	100.0

H-v = High value.

[1]Farms reporting production under production contracts, marketing contracts, or both.

[2]Includes commodities under production or marketing contracts.

[3]The value of production measures the value of commodities produced in a given year, without the effects of inventory change. It is calculated by multiplying the quantity of each commodity produced by the price of the commodity.

[4]The value of production under contract for a given commodity, expressed as a percentage of the group's total value of production under contract.

[5]Ranked in order of largest to smallest, by the commodity's share of the group's contract production. Rankings are based on 10 commodities or commodity groups: grains (including soybeans), cotton, tobacco, high-value crops, other crops, beef, dairy, hogs, poultry, and other livestock.

[6]High-value crops: vegetables, fruits/tree nuts, and nursery/greenhouse products.

[7]Beef and grain were tied for third place at 15 percent.

Source: USDA, National Agricultural Statistics Service and Economic Research Service, 2011 Agricultural Resource Management Survey, Phase III.

Shares of U.S. Totals

Small farms make up a majority (58 percent) of U.S. farms with contracts, and account for 24 percent of farm production under contract, similar to the 23-percent shares for both midsize and large farms. The midsize and large categories, however, make up smaller shares of farms with contracts, 26 percent and 10 percent, respectively. In addition, small, midsize, and large farms each account for about one-fourth of production not under contract, or sold in the cash or spot market. Two commodities make up two-thirds of noncontract production in the United States: cash grain (44 percent) and beef (21 percent), although these commodities may also be contracted.

FARM OPERATORS, PRINCIPAL AND SECONDARY

Every farm has at least one operator, the farmer who makes everyday decisions about the farm business. However, some farms—particularly the larger ones—have more than one operator who makes decisions. In such cases, one operator is designated as the principal operator, the one most responsible for running the farm. The others are considered secondary operators. In the case of single-operator farms, the sole operator is the principal operator.[6]

Commercial-sized farms often require more management and labor than one individual can provide. Additional operators can provide the necessary labor and management and possibly other resources, such as capital or farmland. Having a secondary operator may also provide a successor when an older principal operator phases out of farming.

Secondary Operators and Their Farms

There are secondary operators on 913,800 farms (Table 4). Because farms are generally family businesses, family members usually serve as secondary operators. In fact, 73 percent of the secondary operators—723,700 out of nearly 1 million—are spouses. (The number of secondary operators is 9 percent greater than the number of multiple-operator farms because some multiple-operator farms have more than one secondary operator.)

Table 4. Multiple-operator farms by farm type, 2011

Item	Small family farms				Midsize family farms	Large-scale family farms		Nonfamily farms	All farms
			Farming-occupation						
	Retire-ment	Off-farm occupation	Low-sales	Moderate- sales		Large	Very large		
					Number				
Total operators	488,375	1,330,583	800,621	178,946	194,489	73,023	9,244	96,182	3,171,464
Principal operators[1]	353,922	909,872	567,214	118,253	123,009	38,541	3,857	58,175	2,172,843
Secondary operators	134,453	420,712	233,407	60,694	71,480	34,482	5,387	38,007	998,621
Spouses	103,812	336,644	188,932	41,444	37,669	11,105	1,046	3,067	723,718
Other	30,641	84,068	44,475	19,250	33,811	23,377	4,341	34,940	274,903
Farms with:				*Percent of farms*					
Spouse as an operator	29.3	37.0	33.3	35.0	30.6	28.8	27.1	5.3	33.3
Other secondary operator	7.1	7.4	6.8	13.2	20.4	38.7	54.8	44.2	9.9
Both	0.3	0.9	0.6	2.6	3.5	6.0	9.3	2.7	1.1
				Percent of secondary operators					
Spouse share of secondary operators	77.2	80.0	80.9	68.3	52.7	32.2	19.4	8.1	72.5
				Number					
Operators (principal and secondary per farm)	1.4	1.5	1.4	1.5	1.6	1.9	2.4	1.7	1.5

Table 4. (Continued)

Item	Small family farms				Midsize family farms	Large-scale family farms		Nonfamily farms	All farms
	Retire-ment	Off-farm occupation	Farming-occupation						
			Low-sales	Moderate- sales		Large	Very large		
Multiple-operator farms[2]	128,157	395,378	224,167	53,913 58,432		23,697	2,801	27,261	913,807
				Percent of farms					
Multiple-operator farms as share of all farms	36.2	43.5	39.5	45.6	47.5	61.5	72.6	46.9	42.1

[1]The number of principal operators equals the number of farms. Each farm has one principal operator.

[2]Multiple-operator farms report more than one operator.

Source: USDA, National Agricultural Statistics Service and Economic Research Service, 2011 Agricultural Resource Management Survey, Phase III.

As expected, the number of operators per farm is highest for large-scale farms. The number of operators reaches 2.4 per farm—on average—for very large family farms. The share of family farms with more than one operator also peaks at 73 percent on very large farms, 30 percentage points higher than the share for all U.S. farms.

About 12 percent of all multiple-operator farms (and 5 percent of all farms) are multiple-generation farms, with at least 20 years' difference between the ages of the oldest and youngest operators. Multiple-generation farms are most common among large-scale and nonfamily farms (Figure 6). Principal and secondary operators on multiple-generation nonfamily farms are unlikely to be related to each other. They are more likely to be unrelated managers from different generations. (ARMS does not inquire about the relationships between primary and secondary operators who are not married to each other.)

Principal Operators

One striking characteristic of principal farm operators is their advanced age. About 32 percent of farm operators are at least 65 years old (Table 5). In contrast, only 11 percent of self-employed business owners in nonagricultural industries are that old (USDL/BLS, 2012). Retired operators are the oldest group—as would be expected—with an average age of 70 years, followed by low-sales operators, with an average age of 60 years.

The advanced age of farm operators is understandable, given that the farm is the home for most farmers and that farmers can phase out of farming gradually over a decade or more. Improved health care and advances in farm equipment also have allowed farmers to farm later in life than in previous generations (Mishra et al., 2005, p. 14). The ability to farm later in life also helps explain why the number of older principal operators (691,200) is nearly double the number of retirement farms (353,900).

Trends in Aging

Principal operators' average age has been high for years, and it increased from 55 in the late 1990s to 58 beginning in 2008 and continuing through 2011 (Figure 7).[7]

Principal operators of more commercially oriented farms—those with GCFI of $150,000 or more (measured in constant 2011 dollars)— were 3 to 5 years younger than average for all principal operators in any given year, but their average age still increased from 51 years to 55 years over the 16-year period. The youngest operators on multiple-generation farms had the lowest average age (37 to 40 years), as expected, given how multiple-generation farms are defined. The youngest operators on multiple-operator farms that were not classified as multiple-generation had a higher average age, at least 50 years, in the years examined. Most secondary operators—as explained above—are spouses who tend to have ages similar to the principal operator (Hoppe and Korb, 2013, pp. 30-37).

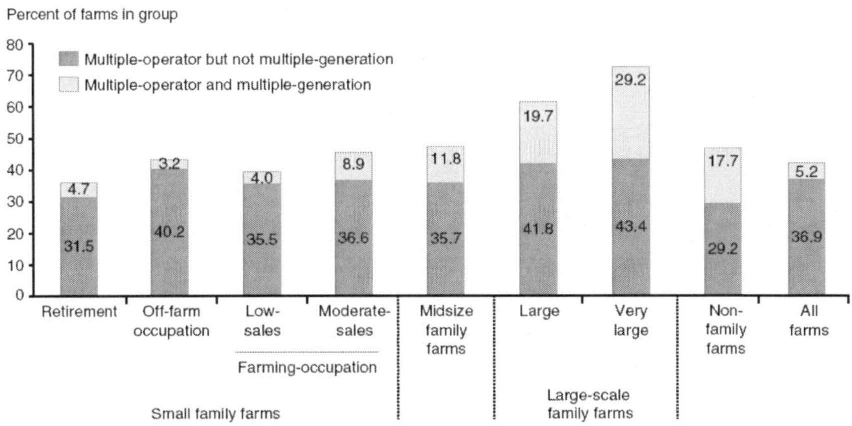

Notes: Multiple-operator farms have more than one operator. Multiple-generation farms are multiple-operator farms with a difference of at least 20 years between the ages of the youngest and oldest operators. The remaining farms are single-operator farms with only one operator (not shown).Source: USDA, National Agricultural Statistics Service and Economic Research Service, 2011 Agricultural Resource Management Survey, Phase III.

Figure 6. Multiple-operator and multiple-generation farms by farm type, 2011Multiple-generation farms are most common among large-scale and nonfamily farms.

Table 5. Selected characteristics of principal operators by farm type, 2011

Item	Small family farms				Midsize family farms	Large-scale family farms		Nonfamily farms	All farms
	Retire-ment	Off-farm occupation	Farming-occupation Low-sales	Farming-occupation Moderate-sales		Large	Very large		
					Number				
Total principal operators	353,922	909,872	567,214	118,253	123,009	38,541	3,857	58,175	2,172,843
					Years				
Average age of principal operator	70	54	60	55	54	54	54	56	58
				Percent of group					
Age of principal operator:									
Younger than 35 years	0.5	4.4	3.9	8.4	5.9	5.5	4.1	4.6	4.0
35 to 44 years	2.6	14.9	5.9	13.5	13.1	13.4	10.2	11.9	10.3
45 to 54 years	4.7	30.1	15.6	23.7	27.0	25.8	31.4	20.6	21.3
55 to 64 years	20.1	33.7	37.0	29.6	36.6	39.1	41.4	43.4	32.6
65 years or older	72.1	16.9	37.6	24.8	17.4	16.2	12.9	19.6	31.8
Education attainment of principal operator:[1]									
Some high school or less	14.4	6.3	11.4	8.5	3.4	2.1	1.9	5.1	8.8
Completed high school	46.0	33.4	52.2	42.4	37.3	39.8	32.5	44.7	41.5
Some college	18.9	23.2	20.6	29.8	33.2	26.2	27.8	23.4	24.9
Completed college	20.7	32.1	15.8	19.3	26.1	32.0	37.8	26.7	24.8

Table 5. (Continued)

Item	Small family farms				Midsize family farms	Large-scale family farms		Nonfamily farms	All farms
			Farming-occupation						
	Retire-ment	Off-farm occupation	Low-sales	Moderate-sales		Large	Very large		
Race of principal operator:									
White	97.7	96.1	94.8	96.7	99.0	98.2	99.2	95.0	96.2
Racial minority[2]	2.3	3.9	5.2	3.3	1.0	1.8	0.8	5.0	3.8
Hispanic origin[3]	5.1	4.0	3.5	2.0	2.0	1.9	3.2	4.4	3.8
Gender of principal operator:									
Male	83.6	90.6	88.3	96.6	97.4	98.8	98.2	92.3	89.8
Female	16.4	9.4	11.7	3.4	2.6	1.2	1.8	7.7	10.2

[1]Vocational school is not counted, unless the credits can be transferred to a college or university. An associate degree is classified as "some college."

[2]American Indians or Alaska Natives, Asians, Blacks or African Americans, and Native Hawaiians or other Pacific Islanders. Also includes operators who reported more than one race. Small sample size for individual racial minorities prevents separate estimates for each group.

[3]Hispanics may be of any race, but 95 percent reported they were White in 2011.

Source: USDA, National Agricultural Statistics Service and Economic Research Service, 2011 Agricultural Resource Management Survey, Phase III.

Most farms are very small and produce little, but this is particularly true among farms operated by respondents who report that they are retired (Table 1). ERS developed a measure of *midpoint age* to track changes in the age of operators responsible for most farm production. Midpoint age in Figure 7 is a median, the midpoint of the distribution of production sorted by the age of the principal operator. The midpoint age reached 55 years by 2008, which means that half of the value of production is on farms with principal operators older than 55 and half is on farms with principal operators younger than 55. The midpoint age has trended upwards over most of the period examined, from a low of 48 years in 1998, increasing by 7 years.

Fewer farms are involved in the "younger half" of production in 2011—below the 55-year midpoint—than in the "older half" (Figure 8).[8] About one-third of all farms have operators younger than 55, while two-thirds have operators age 55 or more, but each group of farms produces similar shares of production. In part, this reflects the large number of farms with operators at least 65 years old and their low production levels.

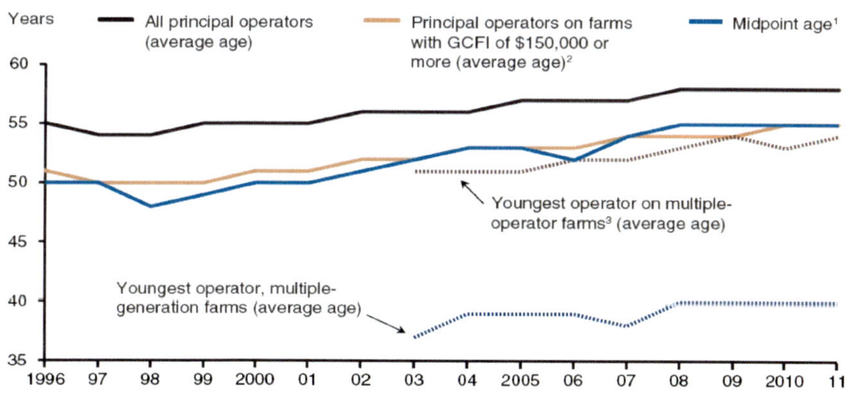

Notes: Both family and nonfamily farms are included in this graph. GCFI = gross cash farm income. 1Half of the value of production is on farms with principal operators older than the midpoint age and half is on farms with principal operators younger than midpoint age. 2GCFI is measured in 2011 constant dollars, using the Producer Price Index (PPI) for farm products to adjust for price changes. 3Excludes multiple-operator farms also classified as multiple generation.Source: USDA, National Agricultural Statistics Service and Economic Research Service, 1996-2011 Agricultural Resource Management Survey, Phase III.

Figure 7. Average age of selected operators and the midpoint age,1 1996 to 2011Age of operators trends upward and half of production now comes from farms with principal operators older than 55 years.

Comparing the 2011 and 1996 distributions of farms and production can provide clues to the future. Between 1996 and 2011, the share of farms and production generally declined for farm operators in age groups below 55 years, while the shares increased for farms with operators 55 to 59 years old and (to a lesser extent) 60 to 64 years old. Little change occurred for older operators at least 65 years old, since operators in this age class are often phasing down their farming operations.

If these patterns continue into the future, the share of farms will continue to decline for operators in age groups below 55 years old. Whether their share of production will also decline is less certain, since younger farmers tend to be more productive.

Shares of farms and production for the 55 to 59 age group may decrease, since the group is currently substantially larger than the next youngest group following it. Shares for the 60 to 64 group are likely to increase due to the aging of the large group of farm operators currently aged 55 to 59 years. The share of farmers 65 years old or more also may increase as younger operators grow older, but their share of production is less likely to increase. It will take years for these potential changes to occur; 15 years passed between the two panels in Figure 8.

Education

Historically, farm operators reported lower levels of educational achievement—measured by the high school completion rate—than the U.S. population in general.

This high school educational gap had largely closed by the late 1980s (Bellamy, 1992, p. 37). More current data show similar shares of high school graduates for farm operators and all U.S. heads of household, 91 and 88 percent, respectively (Figure 9). High school graduation, however, is the highest educational attainment for a larger share of farm operators than for all U.S. householders.

Fewer farm operators complete college (25 percent) than do all U.S. householders (32 percent), and the completion rate for operators of most farm types was also less than the 32-percent national rate (see Table 5). College completion was at the national rate or higher for operators of off-farm occupation farms (32 percent), large farms (32) percent, and very large farms (38 percent). Higher education may be advantageous to farmers when competing for off-farm work or running a large, complex farm business.

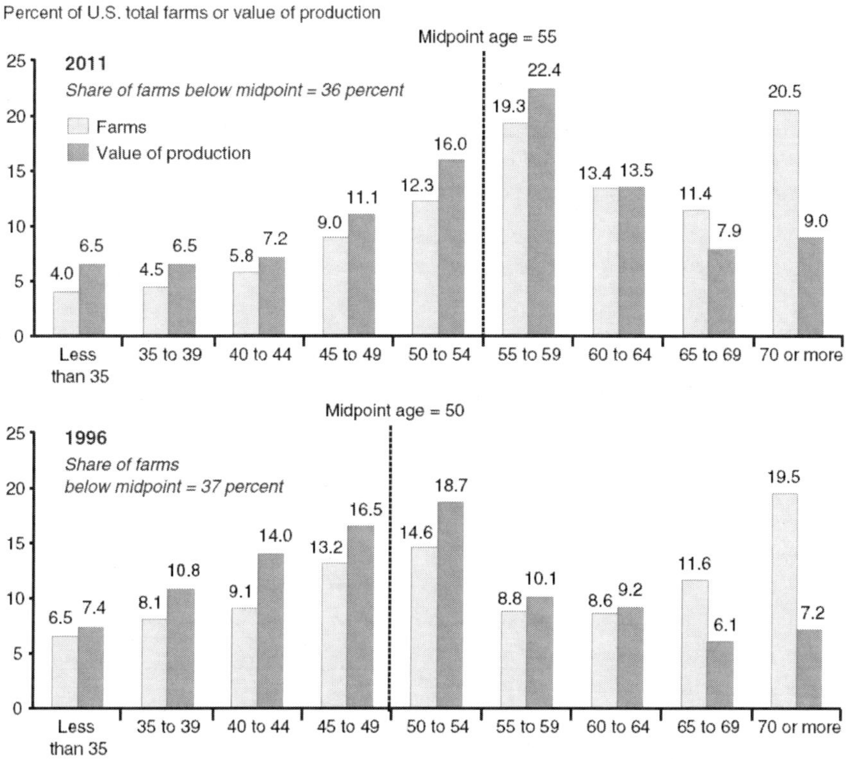

Notes: Both family and nonfamily farms are included in this graph. The value of production measures the value of commodities produced in a year without the effects of inventory change. It is calculated by multiplying the quantity of each commodity produced by the price of the commodity. Source: USDA, National Agricultural Statistics Service and Economic Research Service, 1996 and 2011 Agricultural Resource Management Survey, Phase III.

Figure 8. Distribution of farms and value of production by age of principal operator, 2011 and 1996*The 55- to 59-year age class increased its share of farms and production between 1996 and 2011.*

Gender, Race, and Ethnicity

Principal farm operators are largely White and male. Racial minorities and Hispanics each account for about 4 percent of all principal operators, and similar percentages for each farm type. Men operate 90 percent of all farms and essentially all family farms with GCFI of $150,000 or more (moderate-sales, midsize, large, and very large farms).

Women-operated farms tend to be small; 95 percent are classified as retirement, off-farm occupation, or low-sales farms. The relatively large percentage of retirement and low-sales farms operated by women (16 and 12 percent, respectively) may reflect widows who continue to operate the farm previously operated with their husbands, given the advanced age of principal operators for these farm types. For more information about women farm operators, see *Characteristics of Women Farm Operators and Their Farms* (Hoppe and Korb, 2013).

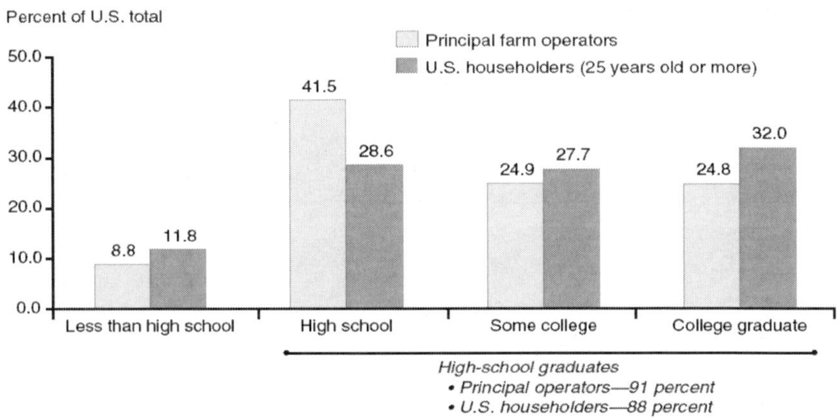

Source: USDA, Economic Research Service and National Agricultural Statistics Service, 2011 Agricultural Resource Management Survey, Phase III, for farm operators; U.S. Census Bureau, Current Population Survey for all U.S. householders.

Figure 9. Educational attainment of principal farm operators and all U.S. householders, 2011*High school is the highest educational attainment for a larger share of farm operators.*

Government Payments and Federal Crop Insurance

The Federal Government supports farmers in various ways, but most directly through programs administered by USDA (Hoppe et al., 2010, p. 23).[9] USDA agencies perform or support agricultural research and extension, provide market information, purchase commodities, and provide services to farmers (such as farm real estate and operating loans and crop insurance). These programs affect farm revenue indirectly by affecting the prices farmers receive, the expenses they pay, or the services they use. Growing ethanol production has also led to increased demand for corn, putting upward pressure on corn prices, but USDA does not administrate ethanol policy.[10] Finally, USDA provides more direct financial support through Government payments from farm programs. This section focuses on two of the principal ways USDA supports farming: Government payments and Federal crop insurance. Both farm programs and Federal crop insurance have existed since the 1930s, helping to support farm income. Crop insurance, however, has grown rapidly in importance in recent years.

Government Payments

Government payments are defined narrowly to include only payments from farm programs to farmers. They exclude payments from Social Security and other public programs, as well as purchases of farm products by State and local governments, such as school districts buying locally produced vegetables for school lunches. Thirty-five percent of farms received Government payments of some sort in 2011, but the relative importance of Government programs was greater for specific farm types (Table 6). Moderate-sales, midsize, and large-scale family farms were more likely to receive Government payments than smaller farms. Government payments—as they existed in 2011, before the 2014 Farm Act—can conveniently be sorted into two groups: commodity-related and conservation (see box, "Types of Farm Program Payments").

Commodity-Related Programs

Commodity programs target specific commodities, largely feed and food grains, cotton, soybeans, other oilseeds, peanuts, and pulses (dry mature peas, lentils, and chickpeas). Payments are tied to the amount of cropland enrolled in programs and yield histories. Specialty (or high-value) crops and livestock production are not generally supported by traditional commodity programs. Producers of nonprogram commodities—as well as producers of program commodities—may also receive disaster assistance and occasional ad hoc payments. Farms specializing in nonprogram commodities may receive substantial payments if they also produce program commodities or did so in the past.

Table 6. Government payments by farm type, 2011

Item	Small family farms				Midsize family farms	Large-scale family farms		Nonfamily farms	All farms
			Farming-occupation						
	Retire-ment	Off-farm occupation	Low-sales	Moderate-sales		Large	Very large		
					Number				
Total farms	353,922	909,872	567,214	118,253	123,009	38,541	3,857	58,175	2,172, 843
				Percent of farms in group					
Farms receiving:									
No Government payments	64.8	76.3	67.6	28.9	18.5	22.0	35.1	64.1	64.9
Any Government payments	35.2	23.7	32.4	71.1	81.5	78.0	64.9	35.9	35.1
Conservation only	21.7	8.3	7.7	4.7	3.0	3.7	6.8	12.2	9.9
Commodity-related only	8.8	12.5	20.1	51.3	54.7	41.7	28.8	15.8	19.0
Both types of payments	4.6	3.0	4.7	15.0	23.8	32.6	29.2	7.9	6.2
				Percent of U.S. total					
Share of payments:									
Total	8.6	10.4	12.1	12.5	27.9	20.7	2.9	4.7	100.0
Conservation	22.3	18.5	17.8	8.3	16.5	10.5	1.9	4.1	100.0
Land retirement	32.6	24.3	19.6	5.6	7.3	6.4	0.5	3.7	100.0
Working land	3.5	8.0	14.5	13.4	33.1	18.1	4.6	4.8	100.0
Commodity-related	2.2	6.6	9.5	14.5	33.3	25.5	3.3	5.1	100.0
Share of:									
Farms	16.3	41.9	26.1	5.4	5.7	1.8	0.2	2.7	100.0
Acres operated	6.5	14.7	17.5	13.4	21.7	14.1	2.1	10.0	100.0
Acres owned by farms	11.6	16.7	19.8	13.6	15.3	8.8	1.3	12.9	100.0
Retired acres[1]	32.4	22.7	19.2	6.0	9.2	6.8	0.5	3.3	100.0

Item	Small family farms				Midsize family farms	Large-scale family farms		Nonfamily farms	All farms
			Farming-occupation						
	Retire-ment	Off-farm occupation	Low-sales	Moderate-sales		Large	Very large		
Acres harvested, selected program crops[2]	1.6	5.5	6.4	12.5	35.6	28.1	4.5	5.7	100.0
					Percent				
Composition of payments:									
Conservation	82.7	56.7	46.8	21.3	18.8	16.3	21.6	27.3	31.9
Land retirement	78.1	48.1	33.3	9.2	5.4	6.4	3.3	16.0	20.6
Working land	4.6	8.5	13.5	12.1	13.4	9.9	18.2	11.4	11.3
Commodity-related	17.3	43.3	53.2	78.7	81.2	83.7	78.4	72.7	68.1
				Percent of land operated					
Share of land enrolled in land-retirement programs on participating farms	47.0	38.4	27.7	9.1	6.6	4.6	2.1	5.1	15.7

[1] Acres that farms enroll in land-retirement programs: Conservation Reserve Program, Conservation Reserve Enhancement Program, and Wetlands Reserve Program.

[2] Corn for grain, cotton, peanuts, rice, sorghum for grain, soybeans, barley for grain, oats for grain, all types of wheat, and canola.

Source: USDA, National Agricultural Statistics Service and Economic Research Service, 2011 Agricultural Resource Management Survey, Phase III.

Types of Farm Program Payments

The 2011 Agricultural Resource Management Survey (ARMS) collected information about the following farm program payments:

Commodity-related payments. Payments from the Direct Counter-cyclical Payment (DCP) and Average Crop Revenue Election (ACRE) programs, loan deficiency payments, marketing loan gains, net value of commodity certificates, milk income loss contract payments, agricultural disaster payments, and any other miscellaneous State, Federal, and local payments. Participation in these programs generally requires present or past production of specific commodities. Goals: Establish price and farm income support, stabilize production, and provide a safety net for farmers.

Conservation payments. There are two types of conservation payments:

- **Payments from land-retirement programs.** Includes the Conservation Reserve Program, the Conservation Reserve Enhancement Program, and the Wetlands Reserve Program (WRP). Goal: Remove environmentally sensitive farmland from production for long periods of time—at least 10 years or permanently, in some cases.
- **Payments from working-land programs.** Includes the Environmental Quality Incentives Program and the Conservation Stewardship Program. These programs provide technical and financial assistance to farmers who install or maintain conservation practices on land in production. **Goal:** Address environmental problems—such as pesticide and nutrient runoff—on land in production.

Federal farm programs listed above were authorized by the **Food, Conservation, and Energy Act of 2008**, which was in effect when the 2011 ARMS was conducted. Since then, Congress passed the **Agricultural Act of 2014,** which eliminates several programs and introduces new ones. In particular, the new act repeals DCP and ACRE programs and introduces the Price Loss Coverage and Agriculture Risk Coverage programs. WRP is consolidated into the new Agricultural Conservation Easement Program.

For more information about these and other changes in the new act, see Agriculture Act of 2014: Highlights and Implications, http://www.ers.usda.gov/agricultural-act-of-2014-highlights-and-implications.aspx

Between two-thirds and three-fourths of moderate-sales, midsize, and large family farms receive commodity-related payments, summing the share receiving only commodity-related payments and the share receiving both commodity-related and conservation payments. These farms collectively received 73 percent of total commodity program benefits paid to farmers in 2011, roughly proportional to their 76-percent share of harvested acres of program crops.

Commodity-related payments in total are much larger than conservation payments, accounting for about two-thirds of all Government payments made to farmers in 2011. Commodity-related payments also make up a majority of Government payments in each farm type, with the exceptions of retirement and off-farm occupation farms.

Conservation Programs

Three USDA land-retirement programs—the Conservation Reserve Program, the Conservation Reserve Enhancement Program, and the Wetlands Reserve Program—together make up 65 percent of all conservation payments paid to farms in the ARMS data. The remaining 35 percent comes from working-land programs: the Environmental Quality Incentives Program and the Conservation Stewardship Program. Historically, land-retirement programs dominated USDA conservation spending, making up about 90 percent of spending on conservation payments made directly to farmers between 1986 and 2003 (Claassen, 2012, pp. 42-43). Beginning in 2003, spending on working-land programs increased nearly tenfold, increasing their share of total conservation spending. The 2014 Farm Act continues the emphasis on working-land programs.

Midsize farms receive the largest share of working-land payments, 33 percent, while low-sales, moderate-sales, and large farms each received roughly half as much. Working-land programs target environmental problems on land in production, encompassing both crop and livestock practices. Small farms with little production—retirement and off-farm occupation farms—receive small shares of these payments. Only 5 percent of working-land programs go to very large farms, reflecting the small share of farmland they operate (2 percent).

Land-retirement programs target environmentally sensitive land which is removed from production, so the distribution of land-retirement payments differs from those of commodity-related payments or working-land payments. Retirement, off-farm occupation, and low-sales farms received 77 percent of land-retirement payments in 2011, reflecting their large numbers (84 percent of all farms), their large share of farmland (48 percent of the land owned by farms), and their tendency to enroll large shares of their land in land-retirement programs when they do participate. Enrollments in land-retirement programs account for 47 percent of the land operated on participating retirement farms, 38 percent on participating off-farm occupation farms, and 28 percent on participating low-sales farms. In contrast, enrollment ranges from 2 percent to 9 percent for the remaining types of farms.

The main occupation of off-farm occupation operators is nonfarm work, which limits the amount of time they can spend farming. Since acreage enrolled in land-retirement programs requires little labor or capital investment and provides a guaranteed income stream, these farmers may find the programs financially attractive, particularly if their farms are not profitable. Given their age, many retired farmers and older farmers on low-sales operations who have eligible land available to put into conservation uses are willing to do so as they scale back their operations and reduce the size of their crop enterprises.

Federal Crop Insurance

USDA's Risk Management Agency (RMA) helps farmers manage risk through insurance provided by the Federal Crop Insurance Corporation (FCIC), which it administers. Federal crop insurance historically has focused on crops, but policies are now available for pasture, rangeland, and livestock. Twenty approved private-sector insurance carriers sell and service the policies. RMA develops insurance products, develops (or approves) premium rates, reinsures the carriers, reimburses the carriers for their administration and operating costs, and subsidizes premiums that farmers pay (Shields, 2013; USDA/RMA, 2010).

More than 120 crops are currently insurable (as of 2011), but counties where insurance is available differ by crop (Shields, 2013, pp. 2-4). Major crops—such as corn, cotton, soybeans, and wheat—can be insured in most counties where they are grown, but less widely grown crops are insurable only in their primary growing areas.

As a result, planted acres covered by crop insurance vary by crop. For example, the share of planted acres covered is 94 percent for cotton, 84 percent for corn and soybeans, 73 percent for fruits and nuts, and 32 percent for vegetables.

Recent History

The use of Federal crop insurance has grown over the past two decades. By 2013, 295.7 million acres were covered, nearly three times the 101.6 million acres covered in 1989 (Figure 10).

The Crop Insurance Reform Act of 1994 and the Agricultural Risk Protection Act of 2000 increased premium subsidies for farmers, and over time new insurance products have been introduced, which increased participation and the amount of coverage purchased (Glauber, 2012, pp. 438 and 487).[11] Premium subsidies account for a large share of the cost of Federal crop insurance (see box "The Cost of Federal Crop Insurance"). About 62 percent of Federal crop insurance premiums are paid by the Federal Government (Shields, 2013, p. 3).

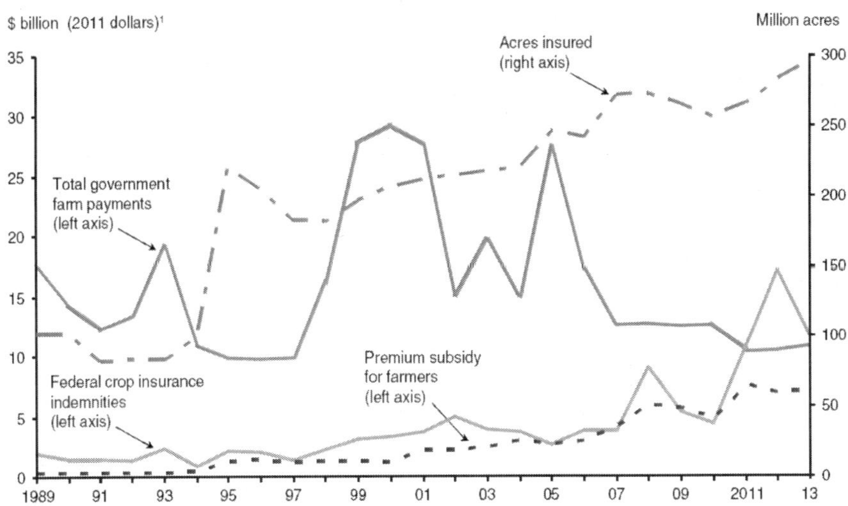

Note: Government payments are reported by calendar year, but crop insurance indemnities, premium subsidies, and acres insured are reported by crop year. A crop year is the 12-month period starting with the month when the harvest of a specific crop typically begins. The 2011 wheat crop year, for example, is June 1, 2011, through May 30, 2012.1Deflated with the gross domestic product (GDP) chain-type price index. Source: USDA, Economic Research Service, U.S. and State Farm Income and Wealth Statistics (the farm sector accounts), available at:

www.ers.usda.gov/data-products/farm-income-and-wealth-statistics.aspx; and USDA, Risk Management Agency, Federal Crop Insurance Corporation, Summary of Business Reports (current as of 06/30/2014), available at: www.rma.usda.gov/data/sob.html.

Figure 10. Government payments and Federal crop insurance, 1989 to 2013Indemnities from Federal crop insurance have exceeded Government payments since 2011.

The Cost of Federal Crop Insurance

The total Government cost of Federal crop insurance was $11.3 billion in fiscal year 2011. Costs of the program consist of claims made in excess of premiums and other income ($2.4 billion), the premium subsidy for farmers ($7.4 billion), expense reimbursements made to the insurance carriers ($1.4 billion), and other costs ($0.1 billion). Premium subsidies have typically made up the largest portion of total costs in recent years. Increasing commodity prices in recent years resulted in higher premiums and premium subsidies.

Cost of Federal crop insurance to the Federal Government, 2011 fiscal year

Cost	Amount	Distribution
	$ million	*Percent*
Claims paid in excess of premiums and other income	2,392	21.2
Premium subsidy paid to farmers	7,376	65.3
A&O expense reimbursements[1]	1,383	12.2
Other program fund costs[2]	144	1.3
Total Government costs	11,295	100.0

[1]Adminstrative and operating (A&O) expense reimbursements are paid to private insurance companies. Premiums reflect only the costs associated with policy risk. The A&O reimbursement pays for delivery costs.

[2]Primarily Federal salaries of RMA personnel and research and development required by the Agricultural Risk Protection Act of 2000.

Claims paid in excess of premiums contribute to the cost of Federal crop insurance. Claims, however, are not necessarily greater than premiums and when claims are less than premiums, the cost of Federal crop insurance is reduced.

Premiums exceeded claims in 7 of the 14 fiscal years between 2000 and 2013.

Note that costs presented here are for fiscal year 2011, but the indemnities and premium subsidies in Figure 10 are for the 2011 crop years for various crops. Costs in a given fiscal year largely reflect the previous crop year, because the fall harvest overlaps the beginning of the fiscal year.

Source: Shields, 2013, pp. 16-17.

Indemnities from Federal crop insurance first exceeded Government payments in 2011 and remained higher than Government payments in 2012 and 2013. Commodity prices have been at historically high levels in recent years, and high commodity prices lower Government payments from programs that are designed to compensate for low prices. But, high prices can increase payments from crop insurance. Indemnities compensate for losses, and the value of losses increase as commodity prices rise.

In addition, adverse weather reduced yields and raised indemnities from 2011 to 2013 (Shields, 2013, p. 11). In 2011, the Great Plains experienced drought in southern and central portions of the region and extensive moisture in the north that hindered planting and production. Indemnities increased again the following year, when a severe drought struck a large part of the country. Continued drought conditions led to high indemnity payments in the Great Plains in 2013.

Farm Participation

About 15 percent of all farms in ARMS reported acres covered by Federal crop insurance, but coverage varied substantially by type of farm (Figure 11). In particular, retirement, off-farm occupation, and low-sales farms—which together account for 84 percent of all U.S. farms—have very low participation rates. The highest participation rates were for large and midsize farms, followed by very large and moderate-sales farms. Farms specializing in cash grains accounted for about 65 percent of all the participants in Federal crop insurance.

Cash grain also was the most common specialization for moderate-sales, midsize, and large farms—accounting for 41, 55, and 48 percent of farms, respectively (see Table 5)—and 77 to 92 percent of cash grain farms in these groups participated in Federal crop insurance. Only 24 percent of very large farms specialized in cash grains, but substantial shares of high-value crops and dairy producers in the group participated in Federal crop insurance.

Seventy percent of U.S. cropland was insured, but coverage was much less for retirement, off-farm occupation, and low-sales farms (Figure 11).[12] In contrast, the more commercially oriented family farms with GCFI of $150,000 or more—as well as nonfamily farms—insured between 62 and 92 percent of their cropland. Overall, farms specializing in cash grains accounted for 68 percent of insured cropland.

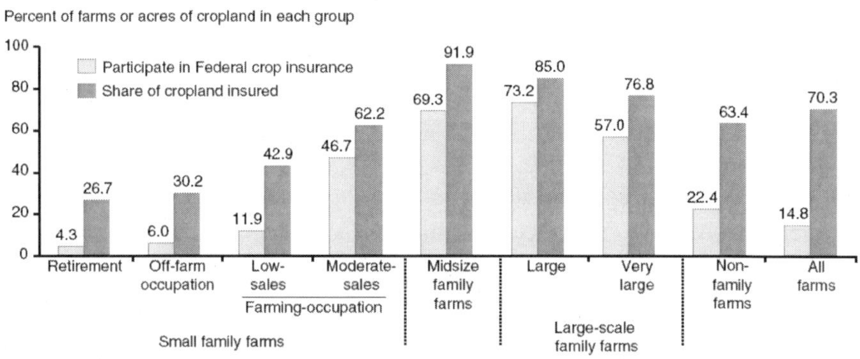

Notes: The ARMS data used here are for calendar year 2011. Cropland excludes any acres enrolled in land-retirement programs. Source: USDA, National Agricultural Statistics Service and Economic Research Service, 2011 Agricultural Resource Management Survey, Phase III.

Figure 11. Participants in Federal crop insurance and the share of cropland insured by farm type, 2011*Midsize and large-scale family farms participate the most.*

The distribution of indemnities is roughly proportional to the distribution of insured acres, as expected (Figure 12). The correspondence is not exact, because indemnities occur only if there is a loss, and the size of indemnities vary with the value of the commodities produced and the level of coverage chosen. Most indemnities (70 percent) accrue to midsize and large farms, reflecting their share of insured cropland acres (66 percent).

Large farms and nonfamily farms each received a small share of indemnities, 6 percent and 4 percent, respectively, reflecting their small share of U.S. farms and insured cropland. Relatively small shares of these farms specialize in grains, a cropland-extensive enterprise. Other specializations common in one or both of the groups—high-value crops, beef (largely feedlots), and dairy—are less cropland extensive.

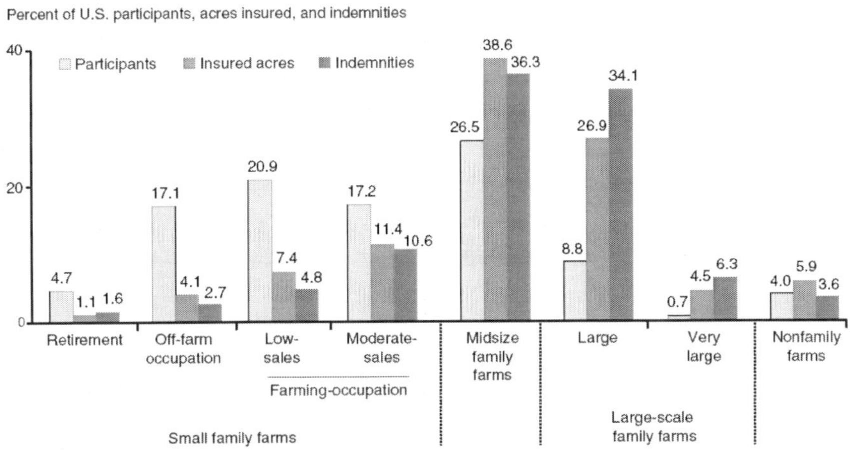

Notes: Indemnities also include those for livestock. The ARMS data used here are for calendar year 2011. Source: USDA, National Agricultural Statistics Service and Economic Research Service, 2011 Agricultural Resource Management Survey, Phase III.

Figure 12. Distribution of Federal crop insurance participants, insured acres, and indemnities by farm type, 2011*Midsize and large family farms receive 70 percent of indemnities, proportional to their share of insured acres.*

FARM INCOME AND FINANCIAL PERFORMANCE

Table 7 presents income statement and balance sheet items as well as selected additional financial measures (see box, "Defining the Financial Measures"). These measures help evaluate farm profitability, financial efficiency, solvency, and financial position, all useful indicators of financial health. Taken together, the measures in Table 7 give a perspective of the financial status of the typical farm and help explain ongoing changes in farms structure. Not all farms perform equally well and performance is generally better for larger farms rather than small farms, although many small farms also perform well.

Earlier editions of the *Family Farm Report* used the mean (or average) as the indicator of the typical level for financial ratios calculated from ARMS, but this edition uses medians. The median of a ratio falls at the midpoint of the distribution of the ratio for farms in a group. Half of the farms have a ratio above the median and half have a ratio below that of the group. The median of the ratio is the ratio for the median farm, the farm at the midpoint of the distribution.

Medians are used because they are affected less by individual cases with extreme values than means. Individual farms may have extremely large ratios, for example, as the denominators approach zero. Means are retained for income statement and balance sheet items, however, because means are additive, which helps show how net farm income and net worth are derived.[13] For more information see "Appendix II: Means and Medians in the ARMS."

Profitability Measures

Three ratios measuring profitability are presented in Table 7:

- **Rate of return on assets** compares profits to resources used to produce them.
- **Rate of return on equity** compares profits to the farm owner's investment in the farm.
- **Operating profit margin** compares profits to the farm's revenue (Northwest Farm Credit Services, 2008, pp. 7-9).

All three profitability measures are strongly associated with farm size. The median rates of return on assets and equity and the median operating profit margin are negative for retirement, off-farm occupation, and low-sales small farms. These measures turn positive, however, for moderate-sales farms, and increase further for midsize and large-scale farms. The ratios return to near zero again for nonfamily farms, reflecting the 78-percent share of those farms with GCFI less than $350,000.

Knowing the median value for a given financial ratio may be useful, but it is also important to know how many farms fall in a critical zone that indicates potential financial problems. Northwest Farm Credit Services (2008) identifies critical zones for 13 financial ratios, including four used in this report (Table 8). Overall, about three-fourths of U.S. farms are in the critical zone for the rate of return on assets, and two-thirds are in the zone for the operating profit margin. The share in the zone for both profitability measures is especially high for retirement, off-farm occupation farms, and low-sales farms, but tapers off rapidly with farm size for moderate-sales and larger farms. A relatively high percentage of nonfamily farms is also in the critical zones for the two ratios, reflecting the small size of most of those farms.

Table 7. Selected financial performance measures by farm type, 2011

Item	Small family farms				Midsize family farms	Large-scale family farms		Nonfamily farms	All farms
			Farming-occupation						
	Retire-ment	Off-farm occupation	Low-sales	Moderate-sales		Large	Very large		
					Number				
Total farms	353,922	909,872	567,214	118,253	123,009	38,541	3,857	58,175	2,172,843
					Percent				
Profitability measures (medians):									
Rate of return on assets	-0.6	-2.9	-2.9	1.1	4.0	8.6	15.3	-0.2	-1.7
Rate of return on equity	-0.7	-3.7	-3.1	0.7	3.7	9.2	18.8	-0.4	-2.2
Operating profit margin	-19.4	-69.0	-53.4	7.1	18.1	24.1	23.8	1.0	-31.2
					Dollars per farm				
Income statement (means):									
Gross farm income[1]	23,532	25,477	41,863	255,197	627,882	2,070,969	10,031,023	867,933	152,642
Less expenses[2]	17,341	23,276	36,619	194,773	480,446	1,529,130	7,248,514	629,053	116,762
Equals net farm income	6,191	2,201	5,244	60,424	147,437	541,840	2,782,510	238,880	35,880
Median net farm income	5,002	788	3,579	67,986	154,538	476,234	1,910,454	6,800	3,631

Table 7. (Continued)

Item	Small family farms				Midsize family farms	Large-scale family farms		Nonfamily farms	All farms
			Farming-occupation						
	Retire-ment	Off-farm occupation	Low-sales	Moderate-sales		Large	Very large		
					Percent				
Financial efficiency measure (median):									
Operating expense ratio	101.4	146.3	103.2	65.9	68.1	65.3	71.0	73.5	99.4
					Dollars per farm				
Balance sheet (means):									
Total assets	627,670	488,737	736,409	1,801,218	2,900,838	6,532,991	17,541,422	3,115,349	991,810
Less total liabilities	10,362	36,554	33,249	162,810	305,918	935,826	3,433,742	244,051	81,082
Equals net worth	617,309	452,183	703,159	1,638,408	2,594,920	5,597,165	14,107,680	2,871,297	910,727
Median net worth	374,206	284,396	442,342	1,188,500	1,918,050	3,633,857	8,335,241	619,870	399,995
					Percent				
Solvency measure (median):									
Debt/asset ratio	0.1	0.2	0.2	4.5	8.4	12.2	21.8	0.2	0.2
Farms with low debt (less than $10,000)	90.9	71.9	73.5	34.6	16.4	7.5	2.7	73.8	69.0
Solvency and income measure:									

Item	Small family farms				Midsize family farms	Large-scale family farms		Nonfamily farms	All farms
			Farming-occupation						
	Retire-ment	Off-farm occupation	Low-sales	Moderate-sales		Large	Very large		
Financial position:									
Favorable	70.6	50.2	60.0	74.3	76.1	73.8	71.5	67.6	59.8
Marginal-income	27.3	41.8	35.9	16.4	16.0	12.8	8.5	29.8	34.1
Marginal-solvency	0.5	3.1	1.7	7.6	5.7	10.7	15.5	1.9	2.8
Vulnerable	1.5	4.9	2.4	1.7	2.2	2.7	4.5	0.7	3.2

Note: Ratios are undefined for individual farms where the denominator is zero or negative, and these cases are excluded when calculating sample medians.

[1]Gross farm income equals gross cash farm income (GCFI) plus net inventory change and nonmoney income (farm household consumption of farm products and the rental value of the farm dwelling). [2]Includes total cash expenses, depreciation, and noncash benefits to labor.

Source: USDA, National Agricultural Statistics Service and Economic Research Service, 2011 Agricultural Resource Management Survey, Phase III.

Defining the Financial Measures

Some of the financial measures used in Table 7 are discussed below. The discussion focuses on the more involved financial measures and shows how they are calculated in the Agricultural Resource Management Survey (ARMS).

Rate of Return on Assets. The ratio of net farm income to the assets of the farm. This ratio is often viewed as an index of profitability, with higher values indicating greater profitability. It is calculated as:

Rate of return on assets = 100 percent × (net farm income + interest paid - charge for operator and unpaid labor – charge for management) ÷ total assets

Interest paid is added back into net farm income because it is the cost of borrowing capital and is part of the return to assets. In the case of unincorporated farms, a charge for operator and unpaid labor and a charge for management are deducted from net farm income to reflect their opportunity cost.

Rate of Return on Equity. The ratio of net farm income to the net worth of the farm. As with the rate of return on assets, it is viewed as an index of profitability, with higher values indicating greater profitability. It is calculated as:

Return on equity = 100 percent × (net farm income - charge for operator and unpaid labor - charge for management) ÷ net worth

Interest paid is not added back into net farm income in this case because the returns to the assets that are owned by the farm are of concern, not those financed through borrowing.

Operating Profit Margin. A measure of profitability: returns per dollar of gross farm income. The operating profit margin measures the funds available to finance the farm business's capital, after accounting for the unpaid labor and management contributed by farm operators and their families. It is calculated as:

Operating profit margin = 100 percent × (net farm income + interest paid - charge for operator and unpaid labor - charge for management) ÷ gross farm income

Operating Expense Ratio. The ratio of cash operating expenses to gross cash farm income. If the ratio is greater than 100 percent, cash income fails to cover cash expenses. It is calculated as:

Operating expense ratio = 100 percent × total cash operating expenses ÷ gross cash farm income

Debt/Asset Ratio. Ratio of the farm's total debt to total assets, showing the share of assets owed to creditors. It is a measure of the risk exposure of the farm business, with a higher ratio corresponding to greater risk.

Debt/asset ratio = 100 percent × total debt ÷ total assets

Financial Position. Measure of the overall financial position of farms based on their combined net income and solvency status:

- **Favorable:** positive net farm income and debt/asset ratio no more than 40 percent.
- **Marginal-income:** negative net farm income and debt/asset ratio no more than 40 percent
- **Marginal-solvency:** positive net farm income and debt/asset ratio greater than 40 percent.
- **Vulnerable:** negative net farm income and debt/asset ratio greater than 40 percent.

Table 8. Farms in the critical zone by farm type, 2011

Item	All farms	Rate of return on assets	Operating profit margin[1]	Operating expense ratio[2]	Debt/asset ratio
			Ratio, expressed as a percent		
Critical zone value[3]	na	< 1.0	< 10.0	> 80.0	> 55.0
	Number		*Percent of farms in critical zone*		
Total farms	2,172,843	74.4	67.3	52.1	3.1
Small family farms:					
Retirement	353,922	74.0	61.8	47.4	1.7
Off-farm occupation	909,872	82.8	73.2	59.1	4.2
Farm-occupation:					
Low-sales	567,214	80.5	74.8	55.6	2.2
Moderate-sales	118,253	49.4	52.4	33.2	3.4
Midsize family farms	123,009	32.2	38.8	30.8	3.0
Large-scale family farms:					
Large	38,541	21.9	29.8	27.5	6.3
Very large	3,857	15.5	25.0	37.2	10.5
Nonfamily farms	58,175	64.9	52.4	37.6	1.8
na = not applicable.					

[1]The 3.8 percent of farms with 0 or negative gross farm income are included in the denominator when calculating the share of farms in the critical zone.

[2]The 15.1 percent of farms with 0 gross cash farm income (GCFI) are included in the denominator when calculating the share of farms in the critical zone.

[3]Critical zones are identified in Northwest Farm Credit Services (2008).

Source: USDA, National Agricultural Statistics Service and Economic Research Service, 2011 Agricultural Resource Management Survey, Phase III.

Differences in profitability also occur among farms within each farm type. For example, 40 to 70 percent of the farms in each small-farm type had a negative operating profit margin in 2011, but other small farms were much more profitable (Figure 13). Between 17 percent and 35 percent of each small-farm type had an operating profit margin of at least 20 percent. Higher profit margins for these small farms slow the shift in production to larger farms (Hoppe et al., 2010, pp. 18-19). Nevertheless, an even greater share of larger family farms had a profit margin of 20 percent or more—roughly half of midsize, large, and very large family farms. Only a minority of farms in these three groups had a negative operating profit margin.

Operating Profit Margin Versus Net Farm Income

Small farms appear more profitable if net farm income is examined rather than the operating profit margin. A majority of each small-farm type generated positive net farm income (Figure 14), but operating profits—the numerator of the operating profit margin—were positive for a majority of farms only in the four groups of family farms with GCFI of $150,000 or more: moderate-sales, midsize, large, and very large farms.

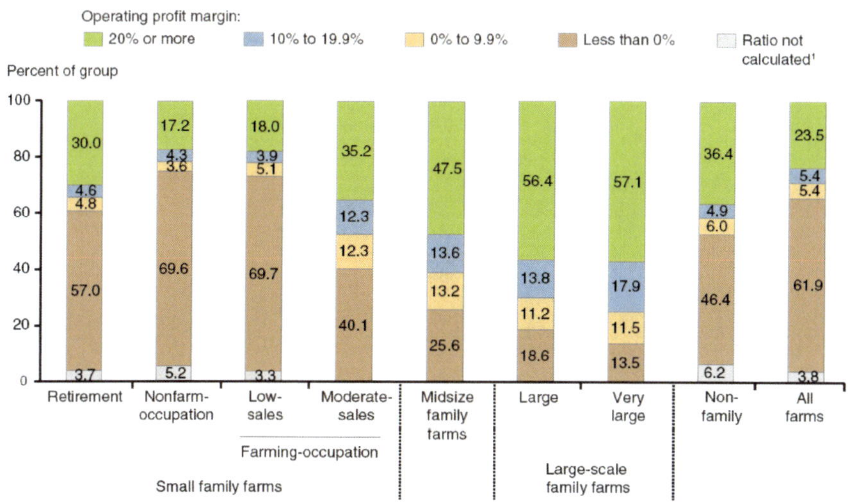

Note: Operating profit = 100 percent X (net farm income + interest paid – charge for operator and unpaid labor – charge for management) ÷ gross farm income. 1The denominator of the ratio—gross farm income—was 0 or negative. Source: USDA, National Agricultural Statistics Service and Economic Research Service, 2011 Agricultural Resource Management Survey, Phase III.

Figure 13. Farms by operating profit margin and farm type, 2011Small family farms are more likely to have a negative operating profit margin than larger family farms.

The different results are attributable to how the two measures treat unpaid labor and management provided by the principal operator and other persons (secondary operators, spouses, and other household members) on unincorporated farms. Operating profit for unincorporated farms is calculated with deductions for unpaid labor and management, to reflect their opportunity cost (Hoppe et al., 2010, p. 17).[14] (See box, "Defining Net Farm Income and Operating Profit.") Net farm income, in contrast, makes no such deductions, placing no value on these resources. Farms with negative operating profits but positive net farm income can stay in business if the operators undervalue their unpaid labor and management.

Both median and mean net farm income increase with farm size and are lower for small farms than for midsize, large, and very large family farms (Table 7). The differences in net income reflect the large differences in gross farm income, from which net farm income is derived. Median net farm income is also low for nonfamily farms ($6,800), reflecting the large share of nonfamily farms with GCFI below the $350,000 cutoff.

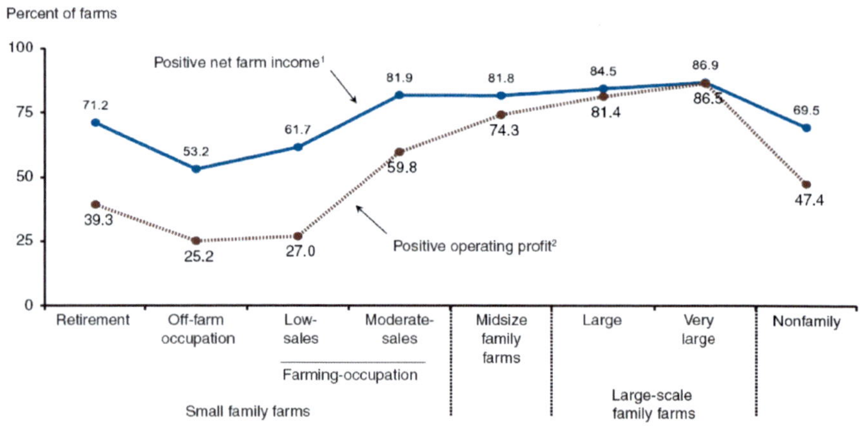

Note: Excludes the 3.8 percent of farms where gross farm income is 0 or negative.₁Net farm income = Gross cash farm income + net inventory change + home consumption + imputed value of farm dwelling – cash expenses – noncash benefits for paid labor – depreciation. ₂Operating profit = Net farm income + interest paid – charge for operator and unpaid labor – charge for management. Source: USDA, Economic Research Service and National Agricultural Statistics Service, 2011 Agricultural Resource Management Survey, Phase III.

Figure 14. Farms by operating profit margin and farm type, 2011Small family farms are more likely to have a negative operating profit margin than larger family farms.

Defining Net Farm Income and Operating Profit

Net farm income =
Gross cash farm income
+ Inventory change
+ Home consumption
+ Imputed value of farm dwelling
- Cash expenses (including interest)
- Noncash benefits for paid labor
- Depreciation
Operating profit =
+ Net farm income
+ Interest paid
- Charge for operator & unpaid labor
- Charge for management

Other Measures

Off-farm occupation farms had a median operating-expense ratio of 146 percent, which means that cash operating expenses exceeded GCFI on the median farm by 46 percent. Households operating off-farm occupation farms generally receive substantial off-farm income to cover living and farm expenses. Cash expenses also exceeded gross cash income for retirement and low-sales farms, but the ratios were lower at 101 and 103 percent, respectively—and near the 100-percent break-even point. The remaining farm types generated enough income to cover expenses, with their median operating expense ratios falling in a fairly narrow range, from 65 to 74 percent.

Half of U.S. farms are in the critical zone for the operating expense ratio (Table 8). This percentage, however, reflects the relatively large shares—47 percent to 59 percent—of retirement, off-farm occupation, and low-sales farms in the critical zone. For the remaining farm types, the share in the critical zone ranges between one-fourth and one-third.

The low median debt/asset ratio for retirement, off-farm occupation, low-sales, and nonfamily farms—all just above 0 percent—reflects their low debt levels (Table 7). At least 70 percent of the farms in each of these groups have less than $10,000 in debt.[15] For the remaining typology groups, the median debt/asset ratio increases with GCFI, from 5 percent for moderate-sales farms to 22 percent for very large farms. Large and very large farms have a relatively large share of farms in the critical zone, two or three times the 3-percent share for all U.S. farms (Table 8). Both large and very large farms carry substantial debt, an average of $0.9 million and $3.4 million, respectively.

ERS developed its financial-position measure in the aftermath of the 1982-1986 farm crisis to sort farms by two measures—net farm income and the debt/asset ratio (Johnson et al., 1987). Most U.S. farms have a favorable financial position, which means they generate positive returns and have a debt/asset ratio of no more than 40 percent. Sixty percent of U.S. farms and at least 50 percent of each farm type were classified as such in 2011 (Table 7). Vulnerable farms—with negative net income and a debt/asset ratio above 40 percent—are rare in all farm types and amount to 3 percent of all farms. Off-farm occupation farms account for 64 percent of the vulnerable farms, but their operators are unlikely to depend on the farm for their livelihood.

OPERATOR HOUSEHOLD INCOME AND NET WORTH

Given their negative operating profit margins and low net farm income—on average—how do so many small farms stay in business? Households operating small farms typically receive substantial off-farm income. In 2011, average off-farm income for small-farm households ranged from $40,600 for moderate-sales households to $97,600 for households operating off-farm occupation farms (Table 9). Most off-farm income (71 percent for all U.S. farm households) is from earned sources, either a wage-or-salary job or self-employment. However, households operating retirement farms receive three-fifths of their off-farm income from unearned sources (such as Social Security, pensions, dividends, interest, and rent), reflecting the advanced age of operators on those farms. (See box, "Measuring Operator Household Income and Net Worth.")

Participation in off-farm work varies by farm type. At one extreme, neither the operator nor spouse worked off-farm on 65 percent of retirement farms. At the other extreme, both the operator and spouse worked off-farm on 56 percent of off-farm occupation farms. In the remaining farm types, someone—the operator and/or the spouse—worked off-farm in 40 to 55 percent of farm households.

OPERATOR HOUSEHOLD NET WORTH

The income that farm operator households receive from farming does not reflect the large net worth of many farm households. As an example, for households on farms with GCFI of at least $150,000, average net worth in 2011 ranged from $1.7 million for moderate-sales farms to $10 million for very large family farms. Unlike operator household income, most of which comes from off-farm sources, net worth from the farm makes up most of the wealth of farm households, regardless of farm type. The farm, on average, accounts for 78 percent of operator household net worth, reflecting the value of the land used in farming. However, much of the net worth of farm households is illiquid and not easily available to spend for consumption because it is largely based on assets necessary to continue farming.

Level of Operator Household Income and Net Worth

Mean income and net worth may not be the best measures of the economic well-being of most farm households, because a few households with high income or net worth can raise the means well above those experienced by the typical farm household. Median household income for all farm households in 2011 was $57,100 (Table 10) or 14 percent higher than the $50,100 median for all U.S. households. Only two types of farm households—those operating retirement or low-sales farms—received household income below the U.S. median. Farm households' net worth compares favorably with that of other U.S. households. For each group of family farms, median net worth was more than the median for all U.S. households ($79,700) or U.S. households reporting a self-employed householder ($294,600) (Bricker et al., 2012). Farm households have substantial farm assets, reflecting the value of their farmland, and most have relatively little farm debt. ERS developed a measure of economic well-being that jointly considers both household income and net worth or wealth (Mishra et al., 2002, pp. 39-42). This measure divides farm households into four groups, based on low and high levels of income and wealth relative to the median household income and wealth for all U.S. households (see Table 10 for the four groups). Roughly 60 percent of households operating retirement or low-sales farm have low income—summing both low-income groups—a larger share than for any other farm types. Nevertheless, most low-income retirement and low-sales households had wealth above the median for all U.S. households.

Table 9. Income and wealth of principal operator households by farm type, 2011

	Small family farms							
			Farming-occupation			Large-scale family farms		
Item	Retire-ment	Off-farm occupation	Low-sales	Moderate-sales	Midsize family farms	Large	Very large	All farm house-holds
				Number				
Total farm households	353,922	909,872	567,214	118,253	123,009	38,541	3,857	2,114,668
				Dollars per household				
Mean household income	64,211	91,748	47,202	84,533	156,417	412,951	1,673,792	87,289
Farm earnings[1]	-755	-5,848	-3,195	43,901	106,834	356,687	1,618,896	14,623
Off-farm income	64,966	97,596	50,397	40,632	49,583	56,264	54,896	72,665
Earned[2]	25,527	82,207	28,165	25,460	34,701	36,969	34,288	51,376
Unearned[2]	39,439	15,390	22,232	15,172	14,882	19,295	20,607	21,289
				Percent				
Share of income from off-farm[3]	101.2	106.4	106.8	48.1	31.7	13.6	3.3	83.2
Off-farm work—by operator and spouse:								
Only operator[4]	12.4	31.6	11.3	7.2	8.9	9.4	8.3	19.8
Only spouse	11.7	1.8	21.1	27.5	32.5	27.9	25.2	12.4
Neither[5]	65.3	11.0[6]	54.4	51.8	45.2	55.1	59.6	36.9
Both	10.6	55.7	13.2	13.5	13.4	7.6	6.8	31.0
				Dollars per household				
Mean household net worth	824,939	680,624	858,388	1,696,336	2,565,255	4,821,053	9,979,909	1,011,309

	Small family farms							
			Farming-occupation			Large-scale family farms		
Item	Retire-ment	Off-farm occupation	Low-sales	Moderate-sales	Midsize family farms	Large	Very large	All farm house-holds
Farm net worth	598,064	435,164	674,628	1,515,363	2,320,188	4,486,010	9,568,248	787,203
				Percent				
Share of net worth from the farm	72.5	63.9	78.6	89.3	90.4	93.1	95.9	7.8

Note: Operator household income and net worth are calculated only for family farms.

[1]Farm earnings in this table and net farm income in Table 7 are not directly comparable. Net farm income includes cash and noncash items and is calculated for the farm business. Farm earnings—in contrast—are based on cash items only, with the exception of a deduction for depreciation. Farm earnings also exclude the share of net income generated by the farm paid to other households, such as those of partners.

[2]Earned income comes from off-farm self-employment or wage or salary jobs. Unearned income includes interest and dividends, benefits from Social Security and other public programs, alimony, annuities, net income of estates or trusts, private pensions, regular contributions of persons not living in the household, net rental income from nonfarm properties, and royalties for mineral leases.

[3]Income from off-farm sources is more than 100 percent of total household income if farm earnings are negative. [4]Includes households where the operator works off-farm and there is no spouse. [5]Includes households where the operator does not work off-farm and there is no spouse.

[6]The off-farm occupation group includes any small farm where the principal operators reports they are not currently in the paid workforce, as well as farms where the operator reports an off-farm occupation. On 13 percent of off-farm occupation farms, the principal operator is not currently in the paid workforce and the spouse also does not work off-farm.

Source: USDA, National Agricultural Statistics Service and Economic Research Service, 2011 Agricultural Resource Management Survey, Phase III.

Measuring Operator Household Income and Net Worth

Operator household income measures the income available to the household of the principal operator. It includes any income received by household members. As measured in the Agricultural Resource Management Survey (ARMS), it has three components:

- Farm business income. In the case of unincorporated businesses and S-corporations, the household's farm business income is calculated as its share of net cash income generated by the farm. Net cash income is gross cash farm income (GCFI)—the sum of the sales of commodities, other miscellaneous farm-related income, and Government payments—less cash expenses and depreciation.* The household of the principal operator does not necessarily receive all the business income generated by its farm. For example, business income may be shared with partners or relatives who hold an interest in the farm. In the case of C-corporations, farm business income is the dividends paid to household members. Wages paid to the operator by farms organized as S- or C-corporations are also included in farm business income.
- Income from other farming activities. This component consists of net income from a farm other than the one being surveyed plus wages paid to household members other than the operator and net income from renting out farmland not associated with the farm being surveyed.
- Off-farm income. Off-farm income can come from earned sources (such as wages, salaries, and self-employment income) or from unearned sources (such as interest, dividends, and Social Security and other transfer payments).

Farm earnings—the income received from farming—are the sum of the first two components. Farm earnings are not directly comparable with net farm income presented in Table 7. Net farm income includes both cash and noncash items and is calculated for the farm business. In contrast, farm earnings apply to the operator household and are based on cash items only, with the exception of a deduction for depreciation. Farm earnings also exclude the share of net income generated by the farm paid to other households.

Unlike net farm income, operator household income excludes two sources of nonmoney income provided by the farm to the farm household: (1) the imputed rental value of the farm dwelling; and (2) the value of farm production consumed on the farm (food and firewood). Average imputed rent, $7,400 per farm, is substantial, which helps explain why average net farm income is positive but average farm earnings are negative for retirement farms, off-farm occupation farms, and low-sales farms (tables 7 and 9). Home consumption is much less, $170 per farm.

ARMS is also the source of data for estimates of operator households' net worth or wealth. The net worth of farm operator households is defined as the difference between the value of their assets and liabilities. It is calculated as the sum of the operator household's farm net worth and nonfarm net worth. If the net worth of the farm is shared with other households (such as the households of shareholders in a family corporation), only the operator household's share is included.

* Depreciation is not a cash expense, but it is deducted to be consistent with accounting conventions used in the Current Population Survey to estimate U.S. household income.

Fifty-six percent of all farm households had both income and wealth above the corresponding medians for all U.S. households. For most typology groups, the share in the high income-high wealth group was even greater, ranging from 66 to 86 percent. The exceptions were households operating retirement or low-sales farms, where only two-fifths were high income-high wealth.

Limited-Resource Farmers and Their Farms

Despite the large share of farm households with high income and wealth, some farmers have limited means. USDA currently defines limited-resource farms as those with low sales that are operated by households with low income (see box, "Defining Limited-Resource Farms"). Limited-resource farms could be more accurately labeled "low income" rather than "limited resource" since the definition limits household income but has no constraint on farm assets or household wealth.

Table 10. Median operator household income and net worth by farm type, 2011

Item	All households	Median household income	Median household net worth	Joint income-wealth indicator[1,2]			
				Low income- low wealth	Low income- high wealth	High income- low wealth	High income- high wealth
	Number	*Dollars per farm household*			*Percent of farm households in group*		
Total farm households	2,114,668	57,050	597,767	2.0	40.9	1.4	55.7
Small family farms:							
Retirement	353,922	39,145	569,154	3.5	57.8	0.4	38.3
Off-farm occupation	909,872	71,566	476,207	1.8	29.5	2.6	66.1
Farming-occupation:							
Low-sales	567,214	42,319	591,620	2.2	56.4	0.6	40.8
Moderate sales	118,253	81,044	1,279,918	0.3	32.8	0.5	66.5
Midsize family farms	123,009	141,234	2,007,157	0.2	21.8	0.3	77.7
Large-scale family farms:							
Large	38,541	356,965	3,192,751	d	17.3	d	81.2
Very large	3,857	1,097,850	6,384,599	d	12.3	d	85.7
Total U.S. households[2]	121,084,000	50,054	79,700	na	na	na	na

Note: Household income and net worth are calculated only for family farms. d = Data suppressed due to insufficient observations na = Not applicable.

[1]The joint indicator compares income and wealth levels for farm households with the median income and wealth of all U. S. households. For example, farm households in the low income-low wealth group have less income and net worth than the corresponding medians

for all U.S. households. In contrast, farm households in the high income-high wealth group have income and net worth equal to or greater than the corresponding medians for all U.S. households.

[2]The income estimate for all U.S. households is from the Current Population Survey, an annual survey conducted by the U.S. Census Bureau (2012, p. 31). The estimate of net worth for all U.S. households is from the Survey of Consumer Finances (SCF), conducted every 3 years. The 2010 estimate of net worth reported in Brinker et al. (2012) was updated to 2011 by adjusting for price changes between the 2010 and 2011, as measured by the Consumer Price Index for all urban consumers.

Source: USDA, National Agricultural Statistics Service and Economic Research Service, 2011 Agricultural Resource Management Survey, Phase III for farm households; U.S. Census Bureau, 2012 Current Population Survey for income of all U.S. households; Board of Governors of the Federal Reserve System, 2010 Survey of Consumer Finances for net worth of all U.S. households.

Defining Limited-Resource Farms

The USDA-wide definition of limited-resource farms was developed by an interagency committee to provide a consistent definition across all USDA agencies.

The definition uses a $100,000 cutoff for farm sales, but it indexes the cutoff to reflect price changes and applies the cutoff to both the current and previous year. Household income also needs to be below the poverty level, or less than half the county median household income, for the current and previous years. In this report, sales and household income must be low in both 2011 and 2010 to classify a farm as limited-resource.

USDA-wide definition of limited resource farms

Criterion	How measured
Sales	Low sales in both the current and previous year; low sales is defined as less than $100,000 in 2003 and indexed thereafter
Operator household income	Low in both the current and previous year. Income is low if it is less than the poverty level for a family of four with two children *or* if it is less than half the county median household income

Source: U.S. National Archives and Records Administration, *Federal Register*, May 30, 2003, p. 32350.

An asset limitation is not used in the definition because the assets held by farmers are difficult to verify on applications to participate in USDA programs. Instead, the requirement for a second year of low income, which is easier to verify than low assets, is added.

For additional information about defining limited-resource farms, see "Special Feature: Limited-Resource Farms—Who Are They?" in the *Family Farm Report, 2010 Edition* (Hoppe and Banker, 2010, pp. 40-45).

Eleven percent of all farm households operate limited-resource farms, but the percentage is much higher for retirement farms (21 percent) and low-sales farms (17 percent) (Figure 15).

The largest share of U.S. limited-resource farms (44 percent of all limited-resource farms) occurs in the low-sales category.

Compared with other farmers, limited-resource farmers have lower median household income, household net worth, and farm assets (Table 11). Nevertheless, 92 percent of limited-resource households are in the low income-high wealth category and have net worth greater than the median for all U.S. households. Limited-resource households also operate smaller farms, on average, than other farm households, in terms of GCFI and acres. About 57 percent of limited-resource farmers are at least 65 years old. Many of these older operators may have scaled back their operation and begun drawing down their assets.

Limited-resource farmers are more likely than other farmers to be female, a member of a racial minority group, or Hispanic. Nevertheless, most limited-resource farmers (72 percent) are non-Hispanic, White males. A smaller share of limited-resource operators are married, which may reflect widows and widowers among the older operators in the group. The smaller share of married operators also means that limited-resource households are less likely than other farm households to have two potential workers (the operator and spouse) to generate farm and off-farm income.

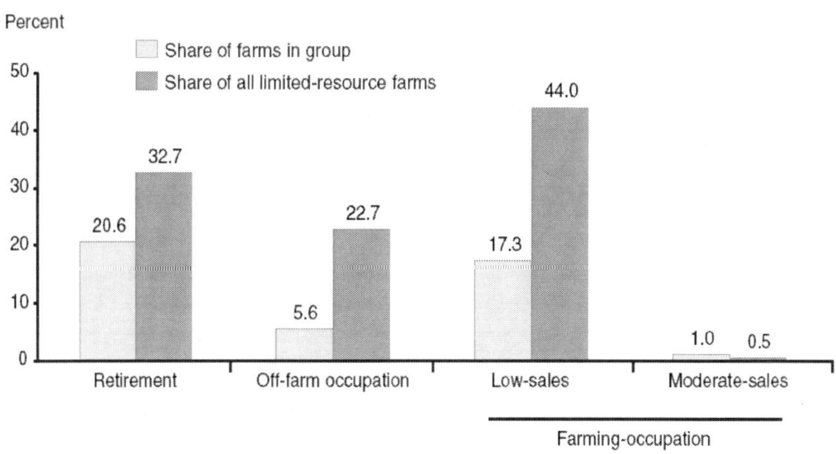

Source: USDA, Economic Research Service and National Agricultural Statistics Service, 2011 Agricultural Resource Management Survey, Phase III.

Figure 15. Limited-resource farms by type of small farm, 2011The largest share (44 percent) of limited-resource farms are low-sales family farms.

Table 11. Selected characteristics of limited-resource and other family farms, 2011

Item	Limited-resource1	Other family farms	All family farms
		Number	
Total farms or households	222,482	1,892,186	2,114,668
		Percent of U.S. total	
Distribution of farms or households	10.5	89.5	100.0
		Dollars per household (or farm)	
Median household income	12,341	64,069	57,050
Median household net worth	391,697	623,700	597,767
Median farm assets	338,612	441,450	427,905
Median gross cash farm income (GCFI)	4,000	8,679	7,800
		Percent of households in group	
Joint income-wealth:			
Low income-low wealth	7.6	1.3	2.0
Low income-high wealth	92.4	34.9	40.9
High income-low wealth	0.0	1.6	1.4
High income-high wealth	0.0	62.2	55.7
		Acres per farm	
Median acres operated	66	85	81
Mean acres operated	166	410	384
		Years	
Average age	65	58	58
		Percent of principal operators	
Operators who are:			
65 years old or more	56.6	29.3	32.2
Female	16.9	9.5	10.3
Racial minority member	6.7	3.4	3.7
Hispanic2	6.8	3.4	3.8
Non-Hispanic, White male	72.1	84.8	83.4
Married	62.8	85.4	83.0
Note: The table excludes nonfamily farms.			

[1]Limited-resource farms are defined as having low sales and low-income operator households. Note that there is no asset constraint in the definition.

[2]Hispanics may be of any race.

Source: USDA, National Agricultural Statistics Service and Economic Research Service, 2011 Agricultural Resource Management Survey, Phase III.

Conclusion and Discussion

This report has four major findings important to understanding farms and farm households, now and in the future:

- Ninety percent of U.S. farms are classified as small family farms, but they only account for 26 percent of production. Most production occurs on midsize and large-scale family farms, although small farms produce substantial shares of specific commodities.
- Midsize and large-scale family farms are generally profitable businesses. Small family farms tend to be less profitable, but the households operating them receive substantial off-farm income to support their farming activities, and do not rely primarily on farming for their livelihoods.
- The financial status of farm households compares favorably with that of U.S. households in general. Except for households operating retirement or low-sales farms, most farm households have higher income and net worth than the median U.S. household.
- The advanced age of principal operators raises concerns about replacements for existing farmers as they age and withdraw from farming. The eventual exit of older farmers, however, is not as ominous as it first appears because farm sector adjustments begin well before older farmers exit.

Midsize and Large-Scale Family Farms Produce the Most

Ninety percent of U.S. farms are small family farms with GCFI less than $350,000, but midsize and large-scale family farms—8 percent of U.S. farms—account for 60 percent of farm production. These larger family farms dominate the production of cotton, cash grain, and hogs. Small farms still account for 26 percent of production, with even higher shares for poultry, hay, other livestock, and beef.

Taken as a group, family farms dominate U.S. agriculture, accounting for 97 percent of farms and 85 percent of production. Several factors favor family farming in the United States (MacDonald, 2014):

- **Technology.** Extensive economies of scale do not exist in farming. Most cost reductions can be attained at a relatively small business size, compared with other industries, even though farming tends to be capital intensive in the United States.

- **Seasonality.** Agriculture can be highly seasonal. Farm households are able to adjust their labor to the seasons by allocating hours from one task to another on the farm or by working off the farm.
- **Local knowledge and expertise.** Crop production requires local knowledge of soils, pests, and weather while livestock production requires knowledge of livestock and how they respond to local conditions. This knowledge takes time to acquire and is not easily transferred to others.
- **Incentives.** Farmers are self-employed business owners and have more incentives to make correct business decisions than do salaried managers. Managers may receive bonuses for good performance or be fired for poor performance, but self-employed business owners reap the increased net worth from well-run enterprises.

Nonfamily farms account for 3 percent of farms and 15 percent of farm production in the United States. But about 85 percent of nonfamily farm production occurs on the 6,300 nonfamily farms with GCFI of $1 million or more, approximately 11 percent of all nonfamily farms. Nonfamily farms with GCFI that high plus large-scale family farms (with GCFI of $1 million or more) together produce about 47 percent of farm output—35 percent on large-scale family farms and 12 percent on nonfamily farms.

Financial Status of the Family Farm

For the most part, midsize and large-scale farms are profitable businesses. Their median operating profit margin and rates of return on assets and equity were all positive, and a large majority of these farms had a positive operating profit margin. In addition, large majorities of farms in these groups were outside the "critical zone" indicating uncomfortably low rates of return on assets and operating profit margins.

In contrast, small farms were less profitable. In the case of retirement, off-farm occupation, and low-sales farms, the median operating profit margin and rates of return on assets and equity were negative. A large majority of the farms in the three groups fell in the critical zone for the rate of return on assets and the operating profit margin. The situation was better for moderate-sales farms. Their medians for the three ratios were positive—but substantially less than those of larger family farms— and roughly half of the farms were outside the critical zone for the rate of return on assets and operating profit margin. Nevertheless, some farms in each small-farm category had a high operating-profit margin, at least 20 percent.

Small-farm households typically receive substantial off-farm income—largely from wage and salary jobs or from self-employment—that they can use to subsidize the farm business and cover living expenses. Because many small-farm households receive a large share of their income from off-farm work, macroeconomic and monetary policies affecting the nonfarm economy are important to them. The provisions in the tax codes allowing farmers to write farm losses off against other income are also important to operators of off-farm occupation farms with substantial off-farm earned income (Durst, 2009, pp.4-5). Finally, the status of retirement programs is important to operators of retirement farms and to older operators in other farm types as they approach retirement.

Financial Status of Farm Households

Most farm households have neither low income nor low wealth. Considering the two measures together, 56 percent of farm households had both income and wealth above the corresponding medians for all U.S. households. For each typology group—except retirement and low-sales farms—a large majority of farm households (66 to 86 percent) were in the high income-high wealth group. In addition, most retirement and low-sales households with low income had wealth above the median for all U.S. households. Much of the wealth of farm households, however, is illiquid and not immediately available for consumption, because it reflects assets necessary for farming.

Even households operating limited-resource farms had high wealth, with 92 percent reporting household net worth greater than the median for all U.S. households. Households operating limited-resource farms are more accurately viewed as having limited income rather than limited resources, since there is no constraint on farm assets or household wealth in the current definition. Assets held by farmers are difficult to verify on applications to USDA programs, so the requirement for a second year of low income—more easily verified than low assets—is used instead. Alternate definitions of limited-resource farms with farm asset or household wealth constraints significantly reduce the number of limited-resource farms (Hoppe and Banker, 2010, p. 48).

Operator household income excludes nonmoney income: the imputed rental value of farm dwellings and the value of products produced on the farm and consumed by the operator household (firewood and food). Exclusion of these items understates the contribution of the farm to the farm operator household's well-being. All farm households receive nonmoney income from the farm, but it may be most important to low-income farm households operating retirement, low-sales, or limited-resource farms.

Older Operators and the Future

The advanced age of U.S. farmers raises concerns about finding replacement farmers as older farmers grow even older and exit farming. The eventual exit of older farmers, however, is not as ominous as it may seem. Substantial numbers of people enter farming, but they are not necessarily young. In 2011, 22 percent of family farms were "beginning farms," operated by farmers with no more than 10 years of farming experience. [16] Only 14 percent of the principal operators of these farms were younger than 35 years. The largest share, 49 percent, was 35 to 54 years old. Beginning farms are more likely than established farms to be small, but both beginning and established farms come in different sizes (Ahearn, 2011 and 2013). Beginning farms are more likely than established farms to be small, but both beginning and established farms come in different sizes (Ahearn, 2011 and 2013).

Secondary operators on multiple-generation farms are another potential source of replacement farmers. While there are relatively few multiple-generation farms—they ranged between 5 percent and 8 percent of all U.S. farms over the years—they are generally much larger than average, and account for a disproportionately large share of agricultural production. Given increasing productivity, fewer—but larger—farms would be necessary in the future to maintain current production levels. Some of the secondary operators on multiple-generation farms would have the experience necessary to operate these larger farms.

Finally, some older operators have already retired and effectively left farming. Operators classified as older are made up of two components: those who are retired (12 percent of all farmers) and those who are not (20 percent). Retired older operators account for only 3 percent of production, and about one-third of their land is either rented to others or enrolled in land-retirement programs.

References

Ahearn, Mary, Jet Yee, Eldon Ball, and Rich Nehring, with contributions by Agapi Somwaru and Rachel Evans. January 1998. *Agricultural Productivity in the United States,* AIB-740, U.S. Department of Agriculture, Economic Research Service.

Ahearn, Mary Clare. 2011. "Potential Challenges for Beginning Farmers and Ranchers," *Choices*, 2nd Quarter 2011.

Ahearn, Mary. January 2013. *Beginning Farmers and Ranchers at a Glance*, EB-22, U.S. Department of Agriculture, Economic Research Service.

Ahrendsen, Bruce L. and Ani L. Katchova. 2012. "Financial ratio analysis using ARMS data," *Agricultural Finance Review*, Vol. 72, Issue 2, pp. 262-272.

Allen, Douglas W., and Dean Lueck. October 1998. "The Nature of the Farm," *Journal of Law and Economics,* Vol. 66, pp. 343-386.

Bellamy, Donald. May 1992. "Educational Attainment of Farm Operators," *Agricultural Income and Finance Situation and Outlook Report,* AFO-45, U.S. Department of Agriculture, Economic Research Service, pp. 37-39.

Bricker, Jesse, Arthur B. Kennickell, Kevin B. Moore, and John Sabelhaus. June 2012. "Changes in U.S. Family Finances from 2007 to 2010: Evidence from the Survey of Consumer Finances," *Federal Reserve Bulletin*, Vol. 98, No. 2, pp. 1-80.

Brown, Jason P., Jeremy G. Weber, and Timothy R. Wojan. November 2013. *Emerging Energy Industries and Rural Growth*, ERR-159, U.S. Department of Agriculture, Economic Research Service.

Cash, A. James, II. December 2002. "Where's the Beef? Small Farms Produce Majority of Cattle," *Agricultural Outlook*, AGO-297, U.S. Department of Agriculture, Economic Research Service, pp. 21-24.

Claassen, Roger. August 2012. "Conservation Spending," in *Agricultural Resources and Environmental Indicators, 2012 Edition*. Craig Osteen, Jessica Gottlieb, and Utpal Vasavada, editors, EIB-98, U.S. Department of Agriculture, Economic Research Service.

Cochrane, Willard W. 1993. *The Development of American Agriculture: A Historical Analysis*. Minneapolis: University of Minnesota Press.

Durst, Ron. May 2009. *Federal Tax Policies and Farm Households*, EIB-54, U.S. Department of Agriculture, Economic Research Service.

Glauber, Joseph W. November 2012. "The Growth of the Federal Crop Insurance Program, 1990- 2011," *American Journal of Agricultural Economics*, Vol. 95, No. 2, pp. 482-288.

Fuglie, Keith O., James M. MacDonald, and Eldon Ball. September 2007. *Productivity Growth in U.S. Agriculture*, EB-9, U.S. Department of Agriculture, Economic Research Service.

Hoppe, Robert A. May 2008. *Structural and Financial Characteristics of U.S. Farms: 2001 Family Farm Report*, AIB-768, U.S. Department of Agriculture, Economic Research Service.

Hoppe, Robert, Penni Korb, and David Banker. December 2008. *Million-Dollar Farms in the New Century*, EIB-42, U.S. Department of Agriculture, Economic Research Service.

Hoppe, Robert A., James M. MacDonald, and Penni Korb. February 2010. *Small Farms in the United States: Persistence Under Pressure*. EIB-63. U.S. Department of Agriculture, Economic Research Service. February 2010.

Hoppe, Robert A., and David E. Banker. July 2010. *Structure and Finances of U.S. Farms: Family Farm Report, 2010 Edition*, EIB-66, U.S. Department of Agriculture, Economic Research Service.

Hoppe, Robert A., and James MacDonald. April 2013. *Updating the ERS Farm Typology*, EIB-110, U.S. Department of Agriculture, Economic Research Service.

Hoppe, Robert A., and Penni Korb. April 2013. *Characteristics of Women Farm Operators and Their Farms*, EIB-111, U.S. Department of Agriculture, Economic Research Service.

Johnson, Cheryl D. May 1990. *A Historical Look at Farm Income*, SB-807, U.S. Department of Agriculture, Economic Research Service.

Johnson, James D., Mitchell J. Morehart, and Kenneth Erickson. 1987. "The Financial Condition of the Agricultural Sector," *Agricultural Finance Review*, Vol. 47, Special Issue, pp. 1-18.

Landau, Zev. November 2005. "Recent Reform and Simplifications for S Corporations," *The CPA Journal*.

MacDonald, James, and Penni Korb. February 2011. *Agricultural Contracting Update, 2008,* EIB-72, U.S. Department of Agriculture, Economic Research Service.

MacDonald, James M., Penni Korb, and Robert A. Hoppe. August 2013. *Farm Size and the Organization of U.S. Crop Farming*, ERR-152. U.S. Department of Agriculture, Economic Research Service.

MacDonald, James. March 2014. "Family Farming in the United States," *Amber Waves*, U.S. Department of Agriculture, Economic Research Service.

McBride, William D., and Kenneth Mathews, Jr. March 2011. *The Diverse Structure and Organization of U.S. Beef Cow-Calf Farms*, EIB-73, U.S. Department of Agriculture, Economic Research Service.

Mishra, Ashok K., Hisham S. El-Ostra, Mitchell J. Morehart, James D. Johnson, and Jeffery W. Hopkins. July 2002. *Income, Wealth, and the Economic Well-Being of Farm Households*, AER-812, U.S. Department of Agriculture, Economic Research Service

Mishra, Ashok K., Ron L. Durst, and Hisham S. El-Osta. April 2005. "How Do U.S. Farmers Plan for Retirement?" *Amber Waves*, U.S. Department of Agriculture, Economic Research Service, Vol. 3, Issue 2, pp. 12-18.

Nickerson, Cynthia, Robert Ebel, Allison Borchers, and Fernando Carriazo. December 2011. *Major Uses of Land in the United States, 2007*, EIB-89, U.S. Department of Agriculture, Economic Research Service.

Northwest Farm Credit Services. 2008. *Understanding Key Financial Ratios and Benchmarks.* Spokane, WA: Business Tools publication.

O'Donoghue, Erik, Robert A. Hoppe, David E. Banker, and Penni Korb. March 2009. *Exploring Alternative Farm Definitions: Implications for Agricultural Statistics and Program Eligibility*, EIB-49, U.S. Department of Agriculture, Economic Research Service.

O'Donoghue, Erik J., Robert A. Hoppe, David E. Banker, Robert Ebel, Keith Fuglie, Penni Korb, Michal Livingston, Cynthia Nickerson, and Carmen Sandretto. December 2011. *The Changing Organization of U.S. Farming*, EIB-88, U.S. Department of Agriculture, Economic Research Service.

Shields, Dennis A. December 2013. *Federal Crop Insurance: Background,* CRS Report-R40532. Congressional Research Service.

U.S. Census Bureau. August 1973. *1969 Census of Agriculture*. Vol. II, General Report. Chapter 3, Farm Management, Farm Operators.

U.S. Census Bureau. 1975. *Historical Statistics of the United States, Colonial Times to 1970*, Bicentennial Edition.

U.S. Census Bureau. September 2012. *Income, Poverty, and Health Insurance Coverage in the United States: 2011*, P60-243, Current Population Reports.

U.S. Department of Agriculture, Economic Research Service. Updated August 26, 2014. *U.S. and State Farm Income and Wealth Statistics*, Web topic page, http://www.ers.usda.gov/data-products/farm-income-and-wealth-statistics.aspx#27396.

U.S. Department of Agriculture, National Agricultural Statistics Service. 2013a. *Agricultural Statistics 2013*.

U.S. Department of Agriculture, National Agricultural Statistics Service. 2013b. *Farm Labor*.
U.S. Department of Agriculture, National Agricultural Statistics Service. May 2014. *2012 Census of Agriculture*, United States Summary and State Data, Volume 1, Geographic Area Series, Part 51: AC-12-A-51.
U.S. Department of Agriculture, Risk Management Agency. November 2010. *About the Risk Management Agency*. Program Aid 1167-02.
U.S. Department of Labor, Bureau of Labor Statistics. February 2003. "Revisions to the Current Population Survey effective in January 2003," *Employment and Earnings*, Vol. 50, No.2, pp. 4-33.
U.S. Department of Labor, Bureau of Labor Statistics. 2012. Annual average household data, "Table 15—Employed persons in agriculture and nonagricultural industries by age, sex, and class of worker," *Employment and Earnings Online*. http://www.bls.gov/opub/ee/2012/cps/ annavg15_2011.pdf
U.S. National Archives and Records Administration, Office of the Federal Register. May 30, 2003. *Federal Register*. Vol. 68, No. 104.
U.S. Office of the President. March 2014. *Economic Report of the President*. Transmitted to the Congress, together with the Annual Report of the Council of Economic Advisors.

APPENDIX I: LARGE AND SMALL NONFAMILY FARMS

The Economic Research Service defines a family farm as any farm where the majority of the business is owned by the operator and individuals related to the operator, including relatives who do not live with the operator. The remaining farms are nonfamily farms, or any farm where the operator and relatives do not own a majority of the business.

Nonfamily farms include more than large farms operated by publicly held corporations trading on a stock exchange. They also include farms organized as cooperatives, a farm equally owned by two or more unrelated business partners, as well as a farm operated by a hired manager unrelated to the owners. Only 17 percent of nonfamily farms are corporations (appendix Table 1) and 94 percent of these corporations report no more than 10 stockholders (see box, "Stockholder Assumptions"). About 12 percent of nonfamily corporations have a hired manager (appendix Figure 1), but the share reaches 71 percent for farms with gross cash farm income (GCFI) of $10 million or more.

Most production by nonfamily farms (94 percent) comes from farms with GCFI of at least $350,000, the cutoff for distinguishing between small and larger farms (see appendix Table 1). Farms with GCFI of $1 million or more account for 85 percent of nonfamily-farm production: those with GCFI of at least $10 million account for nearly half. Nevertheless, 78 percent of nonfamily farms have GCFI less than $350,000.

Whether a farm with GCFI below the cutoff is a small family farm or a small nonfamily farm depends on the ownership structure of the farm. For example, consider a 320-acre farm in Iowa with 240 acres in corn and 80 acres in soybeans generating $307,800 in GCFI in 2011.[17] If the operator and his—or her—relatives have a majority ownership interest in the farm, it would be classified as a moderate-sales family farm. If the farm is operated by two unrelated partners with equal ownership interests, it is a nonfamily farm, albeit a small one. Similarly, if absentee siblings inherit the farm from their parents and hire an unrelated manager, it is a nonfamily farm.

Only one-fifth of the operators of the smallest nonfamily farms (GCFI less than $10,000) report farming as their primary occupation and the share increases rapidly with farm size. When sales exceed $1 million, virtually all operators report farming as their primary occupation. Few operators are retired, except on farms with GCFI less than $10,000 where 37 percent are retired.

Annual person equivalents of labor range from 0.553 for nonfamily farms with less than $10,000 in GCFI up to 133 for farms with $10 million in GCFI. The operator and spouse account for a substantial share of the labor on nonfamily farms with GCFI less than $1 million, between 32 and 66 percent. Once GCFI exceeds $1 million, however, the share of farm labor provided by the operator and spouse labor declines rapidly.

Hired labor's share of total farm labor peaks at 81 percent on farms with GCFI between $5 million and $9,999,999. In addition to hired labor, farms in the highest income class ($10 million or more) also use contract labor, which accounts for about 36 percent of their labor supply. About 47 percent of $10-million nonfamily farms specialize in high-value crops which (when family and nonfamily farms are considered jointly) used about two-thirds of the total hours of contract labor in 2011.

Appendix Table 1. Selected characteristics of nonfamily farms by GCFI class, 2011

Item	Less than $10,000	$10,000-$149,999	$150,000-$349,999	$350,000-$999,999	$1,000,000-$4,999,999	$5,000,000-$9,999,999	$10,000,000 or more	All non family farms
				Number				
Total nonfamily farms	21,214	20,770	3,394	6,499	4,719	732	848	58,175
			Percent of class					
Incorporated[1]	d	d	23.1	40.9	46.2	d	44.3	16.9
			Percent of U.S. total—for nonfamily farms					
Distribution of:								
Farms	36.5	35.7	5.8	11.2	8.1	1.3	1.5	100.0
Value of production[2]	0.1	3.7	1.8	9.8	23.5	13.4	47.7	100.0
			Percent of class					
Primary occupation of principal operator:[3]								
Farming	21.9	50.0	75.0	89.4	96.3	d	d	50.7
Other[4]	78.1	50.0	25.0	10.6	3.7	d	d	49.3
Principal operator is retired	36.7	8.7	d	4.9	d	d	d	17.2
			Annual person equivalents of labor per farm					
Total person equivalents[5,6]	0.553	1.302	3.326	4.565	12.227	50.095	133.013	4.930

Item	Less than $10,000	$10,000-$149,999	$150,000-$349,999	$350,000-$999,999	$1,000,000-$4,999,999	$5,000,000-$9,999,999	$10,000,000 or more	All non family farms
			Percent of total hours					
Share of hours worked by:[7] Operator & spouse[6]	65.7	57.0	31.9	35.5	12.7	3.1	1.3	16.4
Hired labor	8.4	13.1	40.9	38.4	51.9	81.4	61.4	52.1

GCFI = Gross cash farm income.

d = data suppressed due to insufficient observations.

[1]Includes limited liability companies (LLCs) that elect to file their taxes as corporations.

[2]The value of production measures the value of commodities produced in a given year, without the effects of inventory change. It is calculated by multiplying the quantity of each commodity produced by the price of the commodity.

[3]Primary occupation is defined as the occupation at which operators spend 50 percent or more of their work time. [4]Includes operators reporting an off-farm occupation and operators reporting they are not part of the paid workforce. [5]One annual person equivalent equals 2,000 hours of labor, or 50 weeks per year times 40 hours per week. [6]Includes paid and unpaid hours.

[7]Shares worked by other operators, unpaid workers, and contract labor are not shown separately.

Source: USDA, National Agricultural Statistics Service and Economic Research Service, 2011 Agricultural Resource Management Survey, Phase III.

In other words, compared with other crops—such as grains—high-value crops are highly labor- intensive (also see Hoppe et al., 2008, pp. 28 and 30). Production of these crops also has characteristics that make it feasible to use large amounts of labor effectively (Allen and Lueck, 1998). Production is concentrated on fewer acres, compared with row-crops, which makes labor supervision easier.

In addition, in places like California, several cycles of these crops are grown, which means labor use is more continuous and less seasonal. These factors make production of these crops favorable to large-scale family farms and nonfamily farms of a similar size (see Figure 5).

Stockholder Assumptions

The 2011 Agricultural Resource Management Survey did not collect information about the number of stockholders from farm corporations. It did ask, however, about the number of owners associated with each farm. The discussion in Appendix I assumes that the number of owners equals the number of stockholders in the case of farm corporations.

The census of agriculture has asked whether farms organized as corporations have more than 10 shareholders since the 1969 census (U.S. Census Bureau, 1973, p. 127). This report follows the census convention of using 10 or fewer stockholders to indicate small, closely held corporations. The 10-stockholder cutoff was selected because Subchapter S Corporations (S-corporations) could have no more than 10 stockholders in the 1958 legislation originally establishing that form of business organization.

At that time, organizing as an S-corporation was the only way small businesses could obtain the benefit of incorporation—limited liability—without the disadvantage of double taxation. Over the years, the maximum number of shareholders increased to the current 100, but the 10 stockholder-cutoff continued to be used in the census. For more information about the history of S-corporation legislation, see Landau (2005).

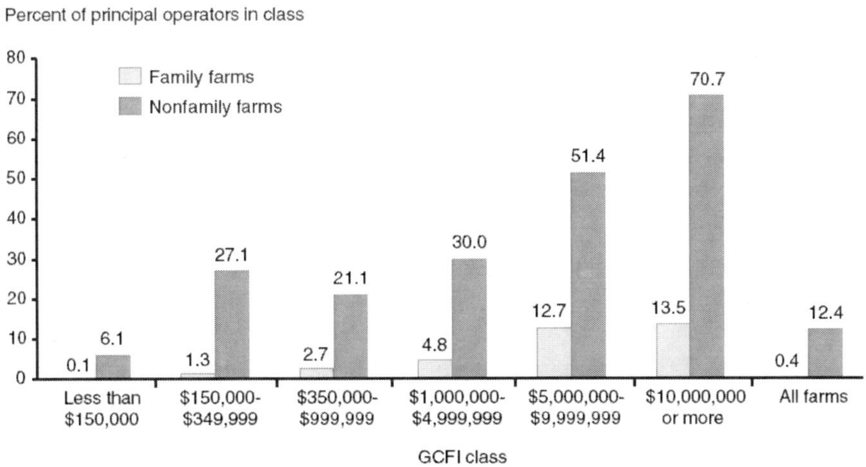

GCFI = Gross cash farm income.

Note: The hired manager of a family farm has an ownership interest in the farm and/or is related to the farm owners. Source: USDA, National Agricultural Statistics Service and Economic Research Service, 2011 Agricultural Resource Management Survey, Phase III.

Appendix Figure 1. Hired managers on family and nonfamily farms by GCFI class, 2011 *Hired managers are more common on nonfamily farms.*

APPENDIX II: MEANS AND MEDIANS IN ARMS

The Economic Research Service (ERS) typically uses aggregate means as a measure of central tendency when calculating financial ratios from the Agricultural Management Resource Survey (ARMS). The aggregate mean of a ratio is calculated as the weighted sum of the numerator for all farms in a given group divided by the weighted sum of the denominator for all farms in the group. As an example, the debt/asset ratio for midsize farms would be calculated as the weighted sum of debts for all midsize farms divided by the weighted sum of their assets. This procedure has the advantage of muting the effect of outliers and avoiding issues related to undefined ratios for individual farms Aggregate means, however, appear to overstate the financial health of the "typical" farm, compared with sample medians (Ahrendsen and Katchova, 2012, pp. 268-271).

Ahrendsen and Katchova (2012) recommend that ARMS analysts consider using sample medians as a measure of central tendency in addition to aggregate means when calculating financial ratios. The sample median is calculated as the weighted median of the ratio for all individual farms in the sample. Sample medians focus on the typical farm which tends to be at the lower end of the size distribution—with lower output—for whatever group is considered. In contrast, aggregate means focus on farm sector performance, reflecting what farms that produce more of the group's output are doing. This appendix explores differences between the two measures.

Appendix Table 2 presents aggregate means for selected financial ratios by type of farm in the first panel and the ratios' corresponding sample medians in the second panel. The third panel shows the difference between the two measures for each ratio and farm type.

The ratios are undefined for individual farms where the denominator is zero or negative, and these cases are excluded when calculating sample medians. This results in dropping 0.5 percent of the cases for the rate of return on equity, 3.8 percent of the cases for the operating profit margin, and 15.1 percent of the cases for the operating expense ratio.

Rate of Return on Assets and Equity

The rate of return on assets calculated as a median is similar—within a percentage point or so—to the rate of return calculated as an aggregate mean for most typology groups. Regardless of how it is calculated, the ratio is small and negative for retirement, off-farm occupation, and low-sales farms, turns slightly positive for moderate-sales farms and then increases with farm size for midsize, large, and very large family farms.

The rate of return on assets, however, is much lower for nonfamily farms when using the sample median (-0.2 percent) than when using aggregate mean (6.8 percent). Although most nonfamily farms—78 percent—are small (GCFI less than $350,000), some are much larger. Eleven percent of nonfamily farms have GCFI of $1 million or more, and they pull the aggregate average rate of return up for the group.

The rate of return on equity follows the same general pattern as the rate of return on assets. Differences between the aggregate mean and sample mean for a given typology group are small except in the case of nonfamily farms. For family farms, the rate of return on equity increases with farm size, just as the rate of return on assets did. The rate of return on both assets and equity for all farms flips from slightly positive when calculated as an aggregate mean to slightly negative when calculated as a sample median.

Appendix Table 2. Aggregate means and sample medians for selected financial ratios by farm type, 2011

Item	Small family farms				Midsize family farms	Large-scale family farms		Non-family farms	All farms
			Farming-occupation						
	Retire-ment	Off-farm occupation	Low-sales	Moderate-sales		Large	Very large		
				Number					
Total farms	353,922	909,872	567,214	118,253	123,009	38,541	3,857	58,175	2,172,843
				Percent					
Aggregate means[1]									
Rate of return on assets	-1.0	-2.2	-2.5	0.8	3.4	7.2	15.1	6.8	1.5
Rate of return on equity	-1.1	-2.8	-2.8	0.2	3.2	7.5	17.4	7.0	1.1
Operating profit margin	-26.8	-41.9	-43.5	5.5	15.9	22.6	26.3	24.6	9.7
Operating expense ratio	98.7	119.4	98.2	72.2	71.7	69.2	69.0	70.6	75.1
Debt/asset ratio	1.7	7.5	4.5	9.0	10.5	14.3	19.6	7.8	8.2
Sample medians[2]									
Rate of return on assets	-0.6	-2.9	-2.9	1.1	4.0	8.6	15.3	-0.2	-1.7
Rate of return on equity[3]	-0.7	-3.7	-3.1	0.7	3.7	9.2	18.8	-0.4	-2.2
Operating profit margin[4]	-19.4	-69.0	-53.4	7.1	18.1	24.1	23.8	1.0	-31.2
Operating expense ratio[5]	101.4	146.3	103.2	65.9	68.1	65.3	71.0	73.5	99.4
Debt/asset ratio	0.1	0.2	0.2	4.5	8.4	12.2	21.8	0.2	0.2

Appendix Table 2. (Continued)

Item	Small family farms				Midsize family farms	Large-scale family farms		Non-family farms	All farms
			Farming-occupation						
	Retire-ment	Off-farm occupation	Low-sales	Moderate-sales		Large	Very large		
				Percentage point difference					
Difference[6]									
Rate of return on assets	0.4	-0.7	-0.4	0.3	0.6	1.4	0.2	-7.0	-3.2
Rate of return on equity	0.4	-0.9	-0.3	0.5	0.5	1.7	1.4	-7.4	-3.3
Operating profit margin	7.4	-27.1	-9.9	1.6	2.2	1.5	-2.5	-23.6	-40.9
Operating expense ratio	2.7	26.9	5.0	-6.3	-3.6	-3.9	2.0	2.9	24.3
Debt/asset ratio	-1.6	-7.3	-4.3	-4.5	-2.1	-2.1	2.2	-7.6	-8.0

[1]Weighted sum of the numerators for all farms in a group divided by the weighted sum of the denominators for all farms in that group. Aggregate means are group-level financial measures.

[2]The weighted median of the ratio for all farms in the group. Sample medians are farm-level financial measures.

[3]Excludes the 0.5 percent of farms with 0 or negative equity.

[4]Excludes the 3.8 percent of farms with 0 or negative gross farm income.

[5]Excludes the 15.1 percent of farms with 0 gross cash farm income (GCFI).

[6]Sample median for a given ratio minus the aggregate mean for the ratio.

Source: USDA, National Agricultural Statistics Service and Economic Research Service, 2011 Agricultural Resource Management Survey, Phase III.

Operating Profit Margin

In contrast, the aggregate mean for the operating profit margin at the all-farms level (10 percent) is much higher than the corresponding estimate based on the sample median (-31 percent). The sample median is more indicative of the experience of most farms, since 62 percent of all U.S. farms reported a negative margin in 2011. The 10-percent aggregate mean reflects high operating profits—the numerator of the ratio—experienced by relatively few farms that mask the extent of negative margins among smaller individual farms.

Switching from the aggregate mean to the sample median also substantially lowers the operating profit margins among off-farm occupation, nonfamily, and low-sales farms—by 27, 24, and 10 percentage points, respectively. Together, off-farm occupation and low-sales farms account for 77 percent of farms with an operating profit margin less than 0, so we would expect margins in these groups to fall after changing from the aggregate mean to the sample median. Likewise, low operating profit margins for the smaller nonfamily farms were masked by high operating profits of larger nonfamily farms when calculating the aggregate mean.

Operating Expense Ratio

Differences in the operating expense ratio by farm type follow the same pattern, whether it is calculated as an aggregate mean or as a sample median. Retirement and low-sales farms are near the break-even level: just above 100 percent (using the sample median) or just below 100 percent (using the aggregate mean). The ratio is substantially higher for off-farm occupation farms, 119 percent if calculated as an aggregate means and 146 if calculated as a sample median. Ratios are less than 100 percent for the remaining farms, ranging from 69 to 72 percent when calculated as an aggregate mean and 66 to 74 percent when calculated as a sample mean.

For most groups, the difference between the operating expense ratios calculated as an aggregate mean or as a sample mean is fairly small, between 2 and 6 percentage points, plus or minus. The exception is the off-farm occupation group, where the ratio calculated as a sample median is 27 percentage points higher when it is calculated as a sample median. Farms in the group tend to have high expenses relative to their revenue, which contributes to their low operating-profit margins as well as their high operating-expense ratios.

Debt/Asset Ratio

Calculating the debt/asset ratio as a sample median rather than as an aggregate mean lowers the ratio from 8 percent to just 0.2 percent at the all-farms level. The small median reflects the small amount of debt held by most farms; 69 percent of all farms have total debt less than $10,000 (see Table 7 in the body of the report). Other farms hold substantial amounts of debt, which results in the 8-percent debt/asset ratio for all farms, when the ratio is calculated as an aggregate mean.

The share of farms with less than $10,000 of debt is high for retirement farms (91 percent), off-farm occupation farms (72 percent), low-sales farms (74 percent), and nonfamily farms (74 percent) (see Table 7). As a result, their debt/asset ratios drop from the 2- to 8-percent range when calculated as an aggregate mean to 0.1 percent or 0.2 percent when calculated as a sample median. Changes for the other farm types were not as dramatic.

End Notes

[1] Cropland must be planted in order to count toward points. In contrast, farms with 100 acres or more of pasture or rangeland but no grazing livestock are classified as point farms (10 points per acre).

[2] ARMS excludes Alaska and Hawaii, largely to reduce the cost of the survey. The 2011 farm counts from ARMS presented in this report are prior to any revisions.

[3] "Other livestock" covers less common livestock species. It also includes grazing livestock other than cattle—horses, sheep, and goats—which are common specializations among small farms.

[4] Point farms (gross sales less than $1,000) make up 25 percent of U.S. farms, or 3 percentage points more than the 22-percent share for farms with GCFI less than $1,000 (Table 1). The 3-percentage point difference has two components: (1) farms with gross farm sales less than $1,000, but GCFI greater than $1,000 (2.8 percent of U.S. farms); and (2) farms with GCFI less than $1,000, but gross farm sales greater than $1,000 (0.2 percent of farms).

[5] The census of agriculture (USDA/NASS, 2014) provides more specializations than ARMS and gives insights into the composition of other crops and other livestock in ARMS. Data from the 2012 Census of Agriculture suggest that about half of other crop farms specialize in hay and four-fifths of other livestock farms specialize in horses, sheep, or goats.

[6] ARMS and the census of agriculture use the same definitions of principal and secondary operators. Starting in 2002, both data sources collect the total number of operators

associated with the farm and ask for detailed information for up to three operators. For more information, see Hoppe and Korb (2013, p. 42).

[7] The analysis in this section is limited to ARMS data, which began in 1996. Data from the census of agriculture, however, indicates that the aging of U.S. farmers has been underway for generations. In the 1940 Census of Agriculture—when the average age of principal operators was first published—operators averaged 48 years of age. The average reached 50 years by 1959, 55 years by 2002, and 58 years by 2012.

[8] Actually, the midpoint age falls *among* principal operators who are 55 years old. Some 55-year-old operators are below the midpoint while others are above it. The first 5 groups' shares of production in Figure 8 total 47.3 percent rather than 50 percent, because *all* the 55-year-olds are in the 55- to 59-year-old category. Similarly, the share of farms operated by principal operators who are younger than the midpoint age is actually 37.0 percent rather than the 35.6-percent sum calculated from Figure 8.

[9] Other Federal activities that support farms include renewable fuel standards, beneficial tax provisions and bankruptcy protection, trade negotiations, investments in public infrastructure, weather service predictions, as well as general business support activities.

[10] Both Federal and State policies increase the demand for ethanol. The 2005 Energy Policy Act, as revised, requires that 15 billion gallons of biofuels be blended with gasoline annually by 2015. Other Federal supports include the Reformulated Gasoline Program and the Winter Oxygenated Fuels Program. State supports include producer incentives, retailer/ infrastructure incentives, State use mandates, retail pump label requirements, and fuel-use requirements for State automotive fleets (Brown et al., 2013, pp 16-17).

[11] The 1994 act also made participation in Federal crop insurance mandatory for farmers taking part in price support programs, production adjustment programs, farm credit, and other farm programs. Opposition to compulsory participation led to its repeal in 1996.

[12] According to RMA administrative data, 83 percent of cropland is insured by Federal crop insurance (Shields, 2013, p. 3), compared with 70 percent in ARMS survey data. The RMA administrative data provide a more complete estimate of insured acres, but ARMS provides detail about individual farms participating in Federal crop insurance. For more information, see White and Hoppe (2012, p. 3). In addition, RMA estimates the share of acres insured using planted acres from NASS as the denominator. ARMS does not collect information on planted acres, so cropland—minus land enrolled in land retirement programs is used instead in Figure 11.

[13] For example, mean assets minus mean liabilities equal mean net worth. Median assets minus median liabilities do not generally equal median net worth.

[14] Unincorporated farms are organized as proprietorships or partnerships. Incorporated farms include both S- and C-corporations. Farms organized as corporations pay salaries to their principal operators. Operators of unincorporated farms receive a nondeductible draw from their farms' equity instead of salaries.

[15] Farm debt in ARMS is the sum of the amount owed on farm loans, accrued interest, and accounts payable. Most farms have accrued interest or accounts payable, but only 31 percent owe on farm loans.

[16] A family farm is considered a beginning farm when a farmer or rancher has not operated a farm or ranch for more than 10 years. This 10-year requirement applies to all operators of the farm or ranch.

[17] The Iowa farm in the example is assumed to have no sources of GCFI other than sales of corn and soybeans. Sales of these commodities were calculated from State-specific yield and price data published in *Agricultural Statistics 2013* (USDA/NASS, 2013a).

In: U.S. Family Farms
Editor: Kathleen Perry
ISBN: 978-1-63483-629-6

Chapter 2

BEGINNING FARMERS AND RANCHERS AT A GLANCE*

Mary Ahearn

Consistent with the declining percentages of beginning farmers and ranchers during the 1982- 2007 period, the share of young operators was also declining (Ahearn and Newton, 2009). In 1982, 16 percent of all principal operators were under 35 years old; by 2007, only 5 percent were under 35 years old. The average age of principal operators of beginning farms in 2011 was 49, compared with 60 for established farms. Even so, only 14 percent of principal operators of beginning farms were under 35 years old. Nearly half of all beginning farmers (49 percent) were 35 to 49 years old, 25 percent were 50-64 years old, and 12 percent were 65 years or older. Most principal operators are men, whether beginning or established, but beginning farms are more likely than established farms to have a principal operator who is a woman (12 percent versus 10 percent).

BEGINNERS LESS LIKELY TO SPECIALIZE

Beginning farms are more likely to be general livestock farms and less likely to specialize in grains and row crops or dairy than established farms. For

* This is an edited, reformatted and augmented version of Economic Brief Number 22, issued by the U.S. Department of Agriculture, Economic Research Service, January 2013.

both types of farms, the most common specialty is beef cattle, but fewer beginning farmers than established farmers specialize in raising cattle.

Beginning farms are less likely to specialize in grains and row crops than established farms

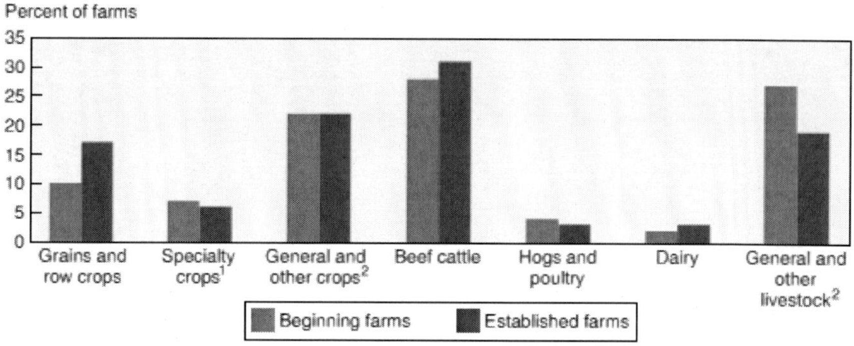

Source: USDA, National Agricultural Statistics Service and Economic Research Service, Agricultural Resource Management Survey, 2011.

[1] Fruits, tree nuts, vegetable, greenhouse, and nursery crops.

[2] General crop and livestock farms are those where no single commodity accounts for 50 percent of the total value of production.

A significant share of U.S. farms produces no agricultural commodities for sale in a typical year. In 2011, 30 percent of beginning farms and 25 percent of established farms did not produce any marketable agricultural commodities. This is largely because most farmers operate small farms, earn significant off-farm income, and/or are elderly. Their property is still defined as a farm, which is any place from which $1,000 or more of agricultural products were produced and sold, or normally would have been sold, during the year (see box, "Definitions").

Both beginning and established farms use a variety of ways to market their products. In recent years, about 10-12 percent of all farms used marketing or production contracts to market their commodities. How products are commonly marketed depends, in part, on what is produced. For example, production contracts are common in poultry and hog production, and marketing contracts are common for many crops, such as sugar beets, peanuts, and some fruits and vegetables. Of farms with positive values of production in 2011, 2 percent of beginning farms used production contracts compared with 3 percent of established farms, and beginning farms also were less likely to use marketing contracts (9 and 14 percent, respectively, in 2011).

Of farms with positive values of production in 2010, 19 percent of beginning farms and 15 percent of established farms sold products directly to individual consumers (for consumption) or to intermediaries who sell directly to individuals (e.g., at a restaurant or grocery store). These beginning farms are more likely than established farms to be engaged in direct sales, such as at farmers' markets (18 percent and 13 percent in 2010), and fewer than 3 percent of beginning farms sold to local retail intermediaries in 2009-10. Beginning farms that sold directly to consumers, rather than to local retail intermediaries, were less likely than other beginning farms to make a profit.

HOUSEHOLDS RECEIVE MOST OF THEIR INCOME FROM OFF-FARM SOURCES

For decades, most farm households—whether beginning or established—received a majority of their household income from off-farm sources. On average, off-farm income accounted for 83 percent of farm operator household cash income in 2011. Most farm families operating small farms have negative net farm income (after depreciation)—that is, in a typical year, they lose money farming. The average farm income of beginning farm households is less than that for established farm households ($1,902 versus $18,119 in 2011), but their households had higher average off-farm incomes than established farm households ($89,015 versus $68,172). Beginning farm households received less farm income and more off-farm income than established farms regardless of farm size.

Principal operators provide a majority of the farm labor—operators of beginning farms worked an average of 1,071 hours per year on their farms in 2011, while operators of estab- 100 lished farms worked 1,344 hours. On average, spouses on established farms also con- 80 tributed more hours of farm work per year (369 in 2011) than those on beginning farms (320 hours). However, while farm families 40 tended to provide fewer total hours on beginning farms, they provided a larger share of 20 the total farm labor hours (87 percent) in 2011, compared with established farms (72 percent). Beginning farms are smaller, on average, than established farms, and larger farms are more likely to hire labor to meet their greater labor requirements.

Households operating beginning farms have higher average off-farm income and lower average farm income at every farm size

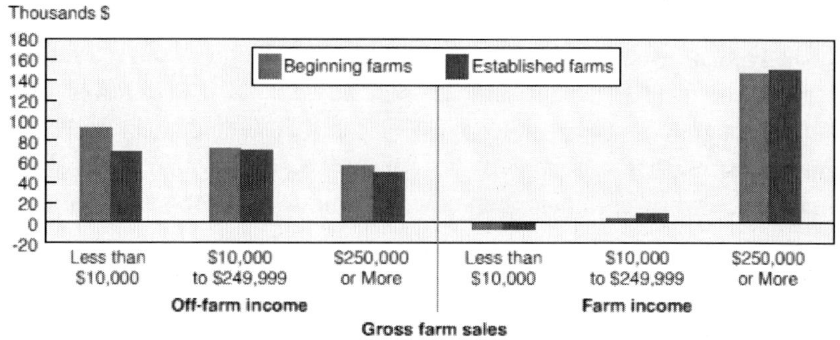

Source: USDA, National Agricultural Statistics Service and Economic Research Service, Agricultural Resource Management Survey, 2011.

Note: Households of principal operators only.

Operators and spouses on beginning farms are more likely to work at off-farm jobs and businesses, and operators of beginning farms are more likely to have a college degree than established farm operators (30 percent compared with 23 percent in 2011). Off-farm income can enable beginning farmers and spouses to meet living expenses and/or expand farm operations if their long-term goal is to farm full time.

Households operating beginning farms work more off the farm, less on the farm

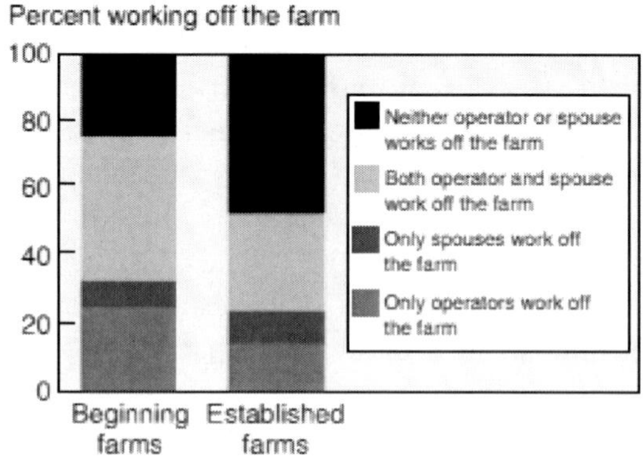

Source: USDA, National Agricultural Statistics Service and Economic Research Service, Agricultural Resource Management Survey, 2011.

Family farm households allocate a disproportionately large share of their time to farming compared with the share of income they receive from farming. This is because returns from farming transcend cash income. For example, the farm often provides tax advantages and long-term capital gains from farmland appreciation. During the last decade, the average annual rate of return to farm assets (a measure of profitability) was 2.1 percent and, when real capital gains are considered, the return was 6.6 percent. Additional returns from farming include onfarm housing and enhanced quality of life.

THE CHALLENGE OF ACQUIRING CAPITAL AND LAND

Beginning farmers often report that their biggest challenge in getting started in farming is access to enough capital and farmland to operate at a size capable of earning a sufficient profit (Shute, 2011). The average beginning farm was 200 acres in 2011, compared with 434 acres for established farms. Not surprisingly, the households of beginning farm operators also have lower farm and nonfarm net worth than the households of established farms. The majority of beginning farms include some owned land, but a higher share of beginning farms than established farms have only rented land. The most common way for U.S. farmers to have acquired "owned land" for their operation is to have purchased it from a nonrelative. This is true for both beginning and established farms. Established farms at all size classes, however, are more likely than beginning farms to have inherited or purchased their owned land from relatives.

The challenges of acquiring land in today's market have likely been exacerbated by the prolonged rise in farm real estate values. The per-acre value of farm real estate in 2012 averaged $2,650, up 10.9 percent from 2011, although there was considerable variation across the Nation (USDA/NASS, 2012). The Northern Plains experienced the largest 2011-12 increase in farm real estate value, at 26.7 percent. The highest State-level real estate values were for some of the Northeastern States, where four States had averages exceeding $10,000 per acre. The Corn Belt region as a whole had the highest average regional value of $5,560 per acre.

Not only should a beginning farmer expect to have low returns in the startup phase, but even established farms must be of a significant size before they are likely to earn positive returns. For example, when size is measured by the gross value of production, it is not until farms are in the $25,000-$30,000 range that the majority of farms have positive net cash returns. Less than one-quarter of beginning farms exceed $25,000 in gross value of production. And the average farm asset base for farms with $25,000 to $30,000 in value of production is over $800,000.

A useful indicator of the ability of a farm to expand by borrowing capital is its debt repayment capacity utilization (DRCU). DRCU is calculated as the ratio of current farm debt relative to the maximum feasible farm debt supportable out of farm income (assuming a 6.0-percent interest rate and a 7-year term). Since it considers only the farm income available to support debt, a DRCU exceeding 100 percent indicates that debt payments must be made by drawing on additional sources, such as off-farm income or household/farm assets. Twenty-three percent of beginning farms had a DRCU exceeding 100 percent in 2011, compared with 13 percent of established farms. The average DRCU of beginning farms in 2011 was 50 percent, compared with 28 percent for established farms.

Most common method of land acquisition is from a nonrelative, but established farms of all sizes are more

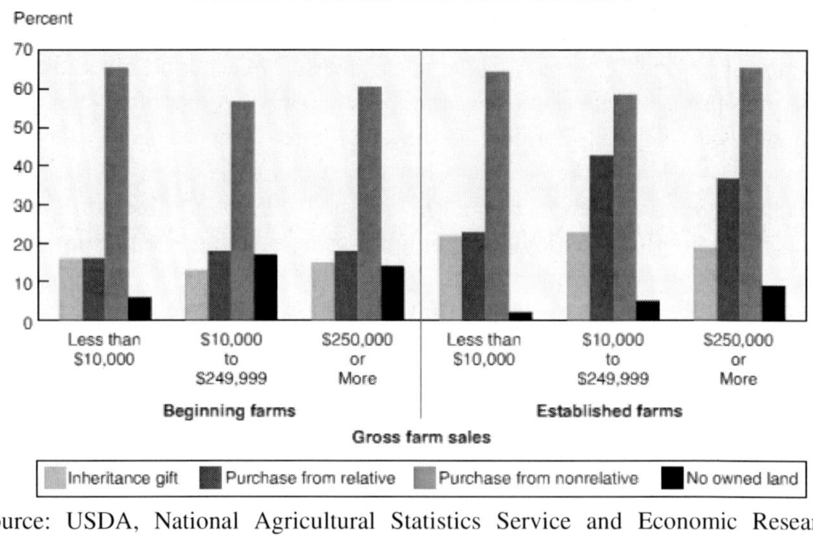

Source: USDA, National Agricultural Statistics Service and Economic Research Service, Agricultural Resource Management Survey, 2010.

GOVERNMENT SUPPORT OF BEGINNING FARMERS

Historically, Government farm programs have targeted specific crops, conservation goals, and disaster relief. But, since 1992, USDA has provided targeted support to beginning farmers and ranchers. Consistent with their production choices, beginning farms are less likely than established farms to participate in Government direct-payment programs, many of which are focused on cash grain production. (Direct payments include those made to farm operators under commodity, conservation, and disaster programs, but exclude loans.) While they represented 22 percent of all farms, beginning farms accounted for only 14 percent of all farms that received Government payments in 2011 and received only 9 percent of all payments that year. Beginning farms are less likely than their established counterparts to participate in Federal crop insurance programs as well; while they accounted for nearly 11 percent of land in farms, they operated only 7 percent of the acres enrolled in crop insurance (USDA, 2011 ARMS).

Beginning farmers and ranchers avail themselves of other farm programs. In fiscal year (FY) 2012 (October 1, 2011 to September 30, 2012), USDA's Farm Service Agency (FSA) made 13,384 direct loans to beginning farmers for a total of $1.1 billion in obligations (almost two-thirds of FSA's total direct loan obligations that year). FSA guaranteed another 2,659 loans to beginning farmers for a total of $639 million, representing 33 percent of FSA's guaranteed loan count and 26 percent of FSA-guaranteed loan obligations in FY 2011. The two programs combined made 50 percent of their loans in FY 2011 and 42 percent of their loan obligations to beginning farmers. In January 2013, FSA announced a microloan program for beginning and small farmers to allow them to borrow small amounts, up to $35,000, for operating loans.

In calendar year 2011, the Farm Credit System (FCS), an independent system of cooperative lenders, made 61,995 loans to beginning farmers, totaling $9.6 billion. These loans represented 19 percent of total FCS farm loans made and 14 percent of total new FCS farm loan volume in 2011.

The 2008 Farm Act authorized $25 million for USDA to establish a Transition Incentive Program (TIP) to encourage enrolled Conservation Reserve Program (CRP) participants to transfer land coming back into production to beginning (and socially disadvantaged) farmers and ranchers engaged in sustainable practices. Enrolled CRP participants must be retired (or farmers or ranchers retiring within 1 year of CRP contract expiration) and are then eligible to receive an additional 2 years of CRP rental payments under TIP. To date, $22.7 million has been allocated. CRP-TIP has 1,719 contracts enrolled or pending enrollment, covering 275,608 acres of land in 26 States.

The 2008 Farm Act also provided for management training programs that target beginning farmers and ranchers. From 2009 to 2012, 145 awards in 46

States, the District of Columbia, and the Virgin Islands were made to training organizations, totaling $70 million.

Beginning farms are less likely to participate in Government direct payment programs, and when they do, average payments are less

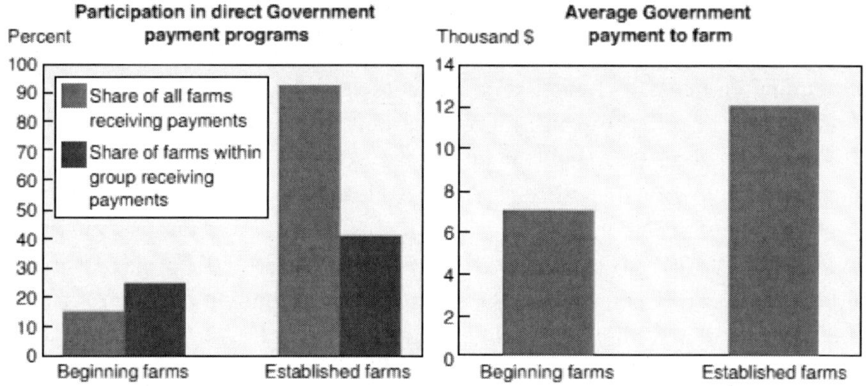

Source: USDA, National Agricultural Statistics Service and Economic Research Service, Agricultural Resource Management Survey, 2011.

Looking to the Future

Beginning farms are experiencing success in a variety of ways, and no one pathway to success is dominant as beginning farmers seek to find their niche. There is some evidence that younger beginning farmers experience higher returns, on average, and are less likely to experience negative returns than older farmers. To the extent that this is true, this is encouraging for the future of U.S. agriculture, since beginning farmers are more likely to be in younger age groups.

Data Sources

The two national sources of data on the beginning-farmer population are the Agricultural Resource Management Survey (ARMS), conducted by USDA's Economic Research Service and National Agricultural Statistics Service, and the Census of Agriculture, conducted every 5 years by NASS (USDA/ERS; USDA/NASS, 2009). See Ahearn and Newton (2009) for a detailed explanation of how these data differ. In the ARMS, the work experience of up to three operators farming their current or other farms is considered. Some farmers have 10 years or less of farming experience and operate a farm jointly with experienced operators; they are not considered beginning farmers and ranchers in this analysis when data on multiple farm operators are available. Of all operators with 10 years or less of farming experience, fewer than 10 percent are jointly operating a farm with an experienced operator.

Data about USDA farm support programs and FCS loans come from several agencies, including:

- Farm Service Agency (FSA). It is estimated that 37 percent of FSA loan guarantees are for Farm Credit System (FCS) loans, so there is some overlap in the FCS loans and FSA loan guarantees reported here. FSA is also a source of data on the Conservation Reserve Program-Transition Incentive Program (CRP-TIP).
- National Institute of Food and Agriculture (NIFA). It provides data on beginning farmer training grants.
- Farm Credit Administration (FCA). Data on FCS lending to beginning farms and ranches are administrative data supplied by FCA. A beginning-farmer loan, as reported by FCA, is a loan made to a farm or ranch where at least one operator has 10 or less years of experience in farming.

Definitions

Farm—Any place from which $1,000 or more of agricultural products were produced and sold, or normally would have been sold, during the year.

Principal operator—The farm operator identified by a survey respondent as being the major day-to-day decision-maker for the farm business. Personal characteristics are reported for the principal operator.

Family farm—The majority of farm assets are owned by the principal operator and his or her extended family. Ninety-seven to 98 percent of U.S. farms are defined as family farms; only 2 to 3 percent are not family farms.

Beginning farm/ranch—All operators have 10 years or less of farming experience.

References

Ahearn, Mary, and Doris Newton. *Beginning Farmers and Ranchers*, EIB-53, U.S. Department of Agriculture, Economic Research Service, May 2009. Available at: http://www.ers.usda.gov/publications/eib-economicinformation-bulletin/eib53.aspx

Shute, Lindsey Lusher. *Building a Future with Farmers: Challenges Faced by Young, American Farmers and a National Strategy to Help Them Succeed*, Tivoli, NY: National Young Farmers Coalition, November 2011.

U.S. Department of Agriculture, Economic Research Service. "Agricultural and Resource Management Survey (ARMS) Briefing Room," May 2012. Available at http://www.ers.usda.gov/Data/ARMS/GlobalAbout.htm

U.S. Department of Agriculture, National Agricultural Statistics Service. *2007 Census of Agriculture*, Geographic Area Series, AC-07-A-51, Vol. 1, February 2009.

U.S. Department of Agriculture, National Agricultural Statistics Service. *Land Values 2012 Summary*, August 2012.

ERS Website and Contact Person

Information on beginning farmers and ranchers can be found on the ERS website at http://www.ers.usda. gov/topics/farm-economy/beginning-disadvantaged-farmers.aspx. For more information, contact Mary Ahearn at mahearn@ers.usda.gov or 202-694-5583.

INDEX

#

A

B

C

T

U

V

W

Y